Energy Use and Conservation Incentives

William H. Cunningham
Sally Cook Lopreato

with the assistance of
Brondel Joseph
Marian Wossum Meriwether
Pablo Rhi-Perez

The Praeger Special Studies program—utilizing the most modern and efficient book production techniques and a selective worldwide distribution network—makes available to the academic, government, and business communities significant, timely research in U.S. and international economic, social, and political development.

Energy Use and Conservation Incentives

A Study of the Southwestern United States

PRAEGER SPECIAL STUDIES IN U.S. ECONOMIC, SOCIAL, AND POLITICAL ISSUES

Praeger Publishers   New York   London

Library of Congress Cataloging in Publication Data

Cunningham, William Hughes.
    Energy use and conservation incentives.

    (Praeger special studies in U.S. economic , social,
and political issues)
    Bibliography: p. 121
    1. Energy consumption—Southwest, Old.   2. Energy—
conservation—Southwest, Old.   I. Lopreato, Sally Cook,
1947-   joint author.   II. Title.
HD9502.U53A1653   1977   333.7   77-7485
ISBN 0-03-022276-1

PRAEGER SPECIAL STUDIES
200 Park Avenue, New York, N.Y., 10017, U.S.A.

Published in the United States of America in 1977
by Praeger Publishers,
A Division of Holt, Rinehart and Winston, CBS, Inc.

789   038   987654321

© 1977 by Praeger Publishers

Printed in the United States of America

ACKNOWLEDGMENTS

The authors would like to express their thanks to the following groups that have, in one way or another, supported the study described in this book: the Center for Energy Studies and the Graduate School of Business, University of Texas at Austin; U.S. Energy Research and Development Administration (Office of Conservation, Division of Buildings and Community Systems, Consumer Motivation and Behavior Branch); and Southern Union Gas Company. Glenn H. Moore assisted greatly in preparation of the survey annotations in Appendix B. We also wish to convey our appreciation to the more than 2,400 people who took the time and care to complete the questionnaire and return it for this analysis.

CONTENTS

Chapter   Page

LIST OF TABLES

LIST OF FIGURES

LIST OF ABBREVIATIONS

BTU   British thermal unit

CEA   Council of Economic Advisors

FEA   Federal Energy Administration

FPC   Federal Power Commission

kwh   kilowatt-hour

mcf   1,000 cubic feet

NORC   National Opinion Research Center

ORC   Opinion Research Corporation

Energy Use and Conservation Incentives

1

THE HISTORY AND
DIFFICULTIES OF
ENERGY CONSERVATION

Recent disruptions of energy supplies to the United States and associated increases in the price of energy fuels have combined to create a situation in which consumers are pressed to reevaluate their energy-use behavior, to adopt more energy-conserving practices, and to purchase more energy-efficient goods. They are pressed from two directions: the one, economic; the other, political. Ever-increasing fuel prices make energy use costly, and have created a situation in which energy can no longer simply be taken for granted. Political pressures to reduce U.S. dependence on foreign energy sources are manifested in programs to approach, if not secure, "energy independence," in part through conservation. Confronted with these pressures, some major questions unresolved for energy policy today are what the consumer thinks about the energy situation, what he will do in terms of energy conservation, and what influences could best effect a move toward more conservation. The present analysis attempts to answer these questions, at least for part of the U.S. public.

THE NATIONAL ENERGY PROBLEM

The sections that follow review briefly the origins of the nation's energy problem. Domestic energy production has declined during the last several years while price has increased and consumption has remained fairly constant. As a result of disparities between domestic consumption and production, the United States faces a growing dependence on foreign energy supplies.

Domestic Production and Consumption of Energy

Annual production of natural gas in the United States went from 6,262 trillion cubic feet in 1950 to 22,600 trillion in 1973, then began a decline, reaching 20,100 trillion in 1975. Crude oil production peaked at 3,577 million barrels in 1971, and then declined to 3,356 million in 1973, and 3,052 million in 1975. Coal production has increased, however, from 560 million short tons in 1950 to 598 million in 1973 and 646 million in 1975 (Council of Economic Advisors [CEA] 1976a, pp. 88-89). More important than these raw figures, however, are the trends in overall production relative to those in consumption, both of which are shown in Figure 1.1. Projected consumption figures through 1985 are based on present trends, without allowance for conservation.

The trend in energy use is clear. Per capita consumption in the United States has risen from 180 million BTU in 1925 to 350 in 1973. The average per capita figure for the world, by contrast, was 23 million BTU in 1925 and 55 million in 1968 (Ford Foundation 1974, p. 74). Comparisons between the average American consumer and any other citizen of the world are rather dramatic, as Figure 1.2 demonstrates.

As per capita energy consumption increased along with electrification, so did inefficiency in energy conversion. The gap between gross and net energy consumed per capita in the United States has risen from approximately 25 million BTU in 1947 to approximately 65 million in 1975 (U.S. Bureau of Mines 1975, p. 19). Technical conversion losses have increased considerably, even since the 1973 "crisis," as consumption of natural gas and electricity generation have given way to oil and coal—both less efficient fuels.

In 1968 residential consumption was more than 19 percent of the national total, with commercial use accounting for 14.4 percent, industrial use 41.2 percent, and transportation 25.2 percent (Ford Foundation 1974, p. 70). By 1973, the household and commercial sectors together consumed 30 percent of the U.S. total, the industrial sector accounted for 39 percent, and the transportation sector had grown to 31 percent (Science Policy Research Division, Library of Congress 1975, p. 53).

Many projections of future consumption have been made, and most of these make some allowance for conservation. The Ford Foundation Energy Policy Project (1974), for instance, estimates a possible savings of 2 quadrillion BTU in residential use in 1985 and of 6 quadrillion in 2000. These estimates are not based on a "belt-tightening" or "do without" view of conservation, but on a "leak-plugging" approach entailing more efficient use of energy through better insulation and construction; the use of heat pumps; more efficient furnaces and air conditioning units; and the use of solar heating and cooling. As a general rule, it is believed that conservation can cut the rate of growth in household consumption by approximately 1 to 2 percent per year (the historic rate is 3.5 percent per year).

FIGURE 1.1

U.S. Production and Consumption of Energy, 1950-85

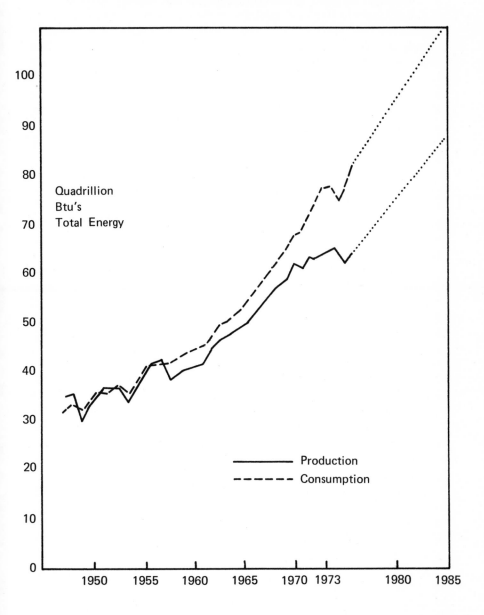

Sources: FEA 1976a, p. 10, 1974b, p. 6; Ford Foundation 1974, p. 71; Science Policy Research Division, Library of Congress 1975, p. 48; U.S. Bureau of Mines 1975, p. 18.

FIGURE 1.2

Energy Use per Capita: United States and World, 1950-68

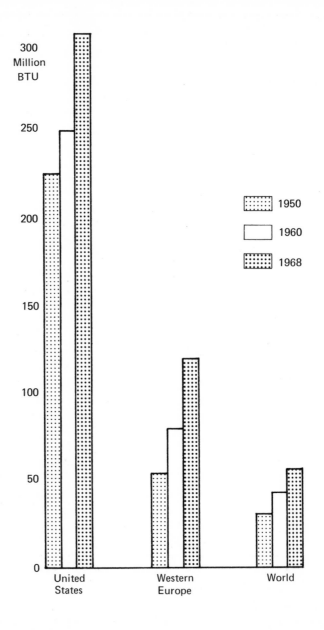

Source: Ford Foundation 1974, p. 73.

4

The growing gap between domestic production and consumption has led to an increased dependence on imports, especially petroleum. Petroleum products supplied 14 percent of the nation's total energy in 1920, but the figure had risen to 46 percent by 1973 (Ford Foundation 1974, p. 69). To meet the petroleum demand, imports of relatively inexpensive crude oil from outside the country, especially Venezuela and the Middle East, increased rapidly. Petroleum imports rose from 8 percent of domestic demand in 1947 to 36.8 percent in 1975, and was expected to reach 46.1 percent in 1977 (FEA 1976a, p. 2) although imports in fact passed the 50 percent mark by 1977. The United States imports roughly 5 percent of its natural gas supply as well (FPC 1976a, p. 1). U.S. dependence on foreign supplies came sharply into focus during the Arab oil embargo in the winter of 1973-74, which created widespread shortages of gasoline and heating fuels.

Prices escalated sharply, then dropped slightly and leveled off, but remained well above pre-embargo levels. The wellhead price of domestically produced crude oil increased 72 percent from 1973 to 1975, and imported crude oil increased even more dramatically. Libyan crude oil, for instance, rose from $3.77 a barrel in January 1973 to $13.71 a barrel in January 1976. Coal prices in 1975 were $16 to $18 a ton free on board (transportation not included) for underground coal and $10 a ton for surface-mined coal, compared with the 1971 composite (average of underground and surface mined) price of $6 a ton (U.S. Bureau of Mines 1976, p. 1). Nationwide, electric utilities paid an average of $18.61 per ton for coal in mid-1976.

The price of natural gas continues to be determined and somewhat confused by the regulatory bodies and long-term contracts that are characteristic of the natural gas business. For 1975 the Federal Power Commission (FPC) set the base ceiling rate to 51 cents per 1,000 cubic feet (MCF) of natural gas that began flowing in interstate commerce after December 31, 1972. This represented an 8-cent increase over the regulated 1974 price. The wellhead price of new gas has since been raised by the FPC to $1.42 per MCF (FPC 1976b). The average price paid for imported gas entering regulated pipelines increased from 59 cents per 1,000 cubic feet in September 1974 to $1.41 a year later and to $1.80 on January 1, 1977 (CEA 1976b). The price of intrastate gas continued to vary widely. For example, in 1976 1,000 cubic feet of natural gas (at retail) cost $3.01 in Austin, Texas, while in Amarillo, Texas, the same quantity of gas could be purchased for 50 cents (TCA Educational Fund 1976).

The Cost of Residential Energy

What all this means to the average American residential consumer is demonstrated in Figure 1.3. The high cost of producing energy has been passed down rapidly to the consumer. Prices of all fuels have risen dramatically. The

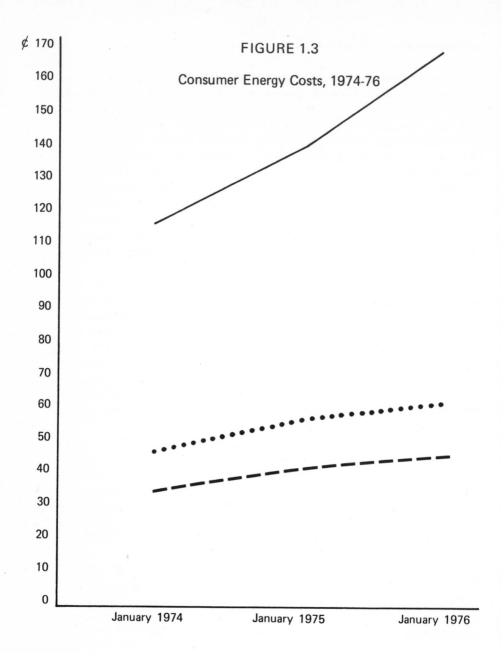

FIGURE 1.3

Consumer Energy Costs, 1974-76

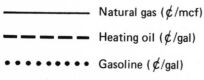

—————— Natural gas (¢/mcf)

— — — — Heating oil (¢/gal)

•••••••• Gasoline (¢/gal)

Source: FEA 1976a, pp. 62, 71, 81.

average retail price of natural gas sold to residential customers for heating rose from $1.133 per 1,000 cubic feet in January 1974 to $1.379 in January 1975, and to $1.674 a year later. Residential heating oil rose from 31.1 cents per gallon in January 1974 to 40.1 cents two years later. The average purchase price of automobile gasoline was 46.3 cents per gallon in January 1974 and rose to 57.7 cents from January to September 1974 and has been level since then; (FEA 1976a, p. 62).

There is some evidence that households affected by rising energy costs have reduced their energy use. From March 1975 to March 1976, consumption in the residential and commercial sectors decreased by 3 percent. While gasoline consumption per licensed driver increased steadily at an average rate of 2.4 percent per year for 1964-73, this figure dropped 6.8 percent in 1974 and decreased by another 0.50 percent in 1975. The total use of petroleum by the residential and commercial sectors declined 5.78 percent from 1973 to 1974 and 5.2 percent from 1974 to 1975. The first quarter of 1976, however, shows a 5.5 percent increase over the same period of 1975 (FEA 1976a, p. 52).

SOUTHWESTERN CONSUMERS

Energy consumption, fuel source, and energy allocation within the home vary greatly within the United States. The FPC has defined eight power supply regions, as seen in Figure 1.4. The Southwest (regions 5 and 8) differs from the rest of the United States in terms of yearly net electrical generation and fossil fuel requirements, as can be seen in Table 1.1. Both regions rely heavily on oil and natural gas and use very little coal in comparison with the North and the South (Science Policy Research Division, Library of Congress 1975).

Household consumption also differs from region to region, mainly as a result of heating and cooling requirements. Figure 1.5 shows U.S. and Texas net residential energy consumption by end use. Nationally, space heating is by far the largest energy user in the home (68 percent). Because of the warmer climate in the Southwest, the average consumer uses considerably less energy than his counterpart in other regions of the country. For example, in Austin, Texas, the average residence uses 90,000 BTU per square foot per year, while the average residence in the Washington-Baltimore area consumes 130,000 BTU per square foot per year. Less than half of the Texas resident's household consumption is for heating, and more than 25 percent is used for air conditioning, compared with approximately 2 percent nationally.

Overview of the Study

Data for the present study of southwestern consumers were collected in October 1975, through a mail questionnaire sent to residents in five

FIGURE 1.4

Power Supply Regions as Defined by the FPC

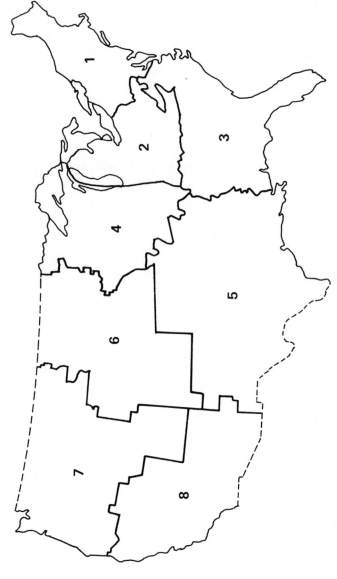

Source: Science Policy Research Division, Library of Congress 1975, p. 31.

TABLE 1.1

Net Electrical Generation by Source, 1973 (million kwh)

	Region I	Region II	Region III	Region IV	Region V	Region VI	Region VII	Region VIII
Coal	102,867	297,010	228,795	131,382	18,731	21,028	12,808	20,639
Oil	159,181	10,655	67,060	6,836	19,737	429	352	51,325
Gas	7,967	5,652	22,994	12,323	198,692	8,549	624	54,229
Other	64,189	6,882	60,016	34,406	895	12,212	117,811	48,943
Total	334,204	320,199	378,865	184,947	238,055	42,218	131,595	175,136

Source: Science Policy Research Division, Library of Congress 1975, pp. 33, 34.

TABLE 1.2

Demographic Characteristics of the Cities Sampled

	Austin	El Paso	Albuquerque	Flagstaff	Prescott
Population	251,817	322,261	243,751	26,117	13,030
Percent male (18 yrs. and over)	48.9	46.1	48.1	49.9	50.1
Percent female (18 yrs. and over)	51.1	53.9	51.9	50.1	54.5
Age distribution					
Percent < 18	31.5	41.1	36.9	33.6	30.1
Percent 18-64	61.6	52.9	56.7	63.1	50.4
Percent > 64	6.9	6.0	6.4	3.3	19.6
Percent unemployed	3.2	5.2	5.3	4.8	5.3
Per capita income	$2,998	$2,390	$3,091	$2,801	$2,687
Percent families below poverty-level income	11.0	16.8	11.1	8.8	14.1
Income distribution					
Median	$9,180	$7,963	$9,641	$9,739	$7,277
Mean	$10,810	$9,469	$10,926	$11,009	$8,609
Racial/ethnic characteristics					
Percent Anglo	72.0	48.0	60.8	71.0	n.a.
Percent Spanish heritage	15.6	58.1	34.9	18.5	n.a.
Percent black and other	12.4	3.9	4.3	10.5	n.a.
Median education (years)	12.5	12.1	12.6	12.6	12.2

n.a. = not available.
Source: U.S. Bureau of the Census 1970, 1972.

FIGURE 1.5

U.S. and Texas Net Residential Consumption by End Use

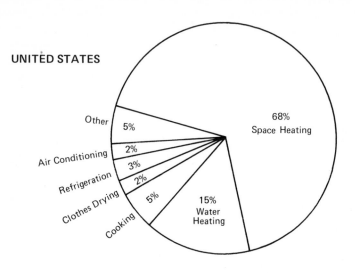

UNITED STATES

68%
Space Heating

Other
5%

2%
Air Conditioning

3%
Refrigeration

2%
Clothes Drying

5%
Cooking

15%
Water
Heating

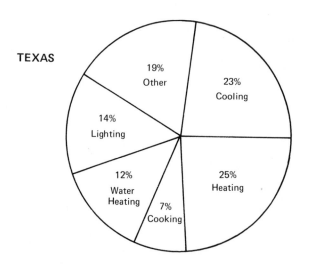

TEXAS

19%
Other

23%
Cooling

14%
Lighting

12%
Water
Heating

7%
Cooking

25%
Heating

Sources: United States–Science Policy Research Division, Library of Congress 1975, p. 59; Texas–Reed 1974, p. 70.

communities: Austin and El Paso, Texas; Flagstaff and Prescott, Arizona; and Albuquerque, New Mexico. These cities were selected to represent various types of communities in the area. The subjects were chosen from the consumer billing records of the local natural gas company. Every tenth residential consumer on the firm's mailing list was included in the sample. Twelve percent of the residential customers serviced by the electric utility in Flagstaff and Prescott are all-electric home customers; 3 percent of the customers of the utility servicing Albuquerque are all-electric customers, as are 9 percent in Austin and 1 percent in El Paso (FPC 1975a). While it cannot be assumed that all of the remaining residents supplement their electrical energy with natural gas, it is highly likely in this region of the country that the majority use some natural gas and are therefore part of the population sampled.

Ten thousand questionnaires were mailed with an accompanying letter explaining that consumers were being asked to participate in a study sponsored by the University of Texas at Austin and that responses would be reported only in aggregate form. A response rate of nearly 25 percent yielded 2,403 codable returns for analysis. This response rate is a good one for mail surveys, particularly given the length and complexity of the questionnaire (see Appendix A).

Basic demographic characteristics of the cities sampled are reported in Table 1.2. El Paso is the largest of the five cities, with a population of more than 300,000 in 1970; Albuquerque and Austin both reported populations of more than 240,000. Flagstaff and Prescott are much smaller, with approximately 26,000 and 13,000 residents, respectively. Twenty-five percent of the sample returns were from each of the three largest cities, with the remainder divided between the two smaller Arizona cities.

Slightly more than 50 percent of the residents of all five cities are female. Persons of Spanish heritage range from 15.6 percent of the population of Austin to 58.1 percent of the population of El Paso. Blacks and other minority races make up 12.4 percent of Austin's population and 3.9 percent of the population of El Paso. El Paso has the youngest population, with more than 41 percent under 18 years of age. Prescott claims the highest percentage of elderly persons, with nearly 20 percent over age 64.

The sample shows a bias toward middle-aged white males with higher than average education and income. More than 70 percent of the respondents were male. Almost 84 percent of the respondents reported their race as white, with only 10.5 percent defining themselves as Mexican-American and 2.4 percent as black. The questionnaire was only in English, and was somewhat long and difficult, which probably contributed to underrepresentation of persons of Spanish heritage. It must be remembered, too, that Mexican-American families tend to be larger than Anglo families (Grebler et al. 1970), so the percentage of Mexican-American households in a given area is smaller than the percentage of individuals in the population. Because the sample is quite large, however, the data include a significant number of persons of both black and Mexican-American minority groups. Thus, trends and patterns in attitudes and behavior can be effectively studied, even though representation of these groups in the sample does not approach overall representation in the population. The majority of the sample fell in middle-age categories, with 7 percent of the sample reporting their age as under 25 and 16 percent over 60.

The average median number of school years completed for the five cities is slightly more than 12. Less than 33 percent of the sample reported no college education, and almost 33 percent reported graduate-level work or degrees. Likewise, incomes of the sample population were relatively high, with almost 50 percent reporting yearly family incomes of $15,000 or above. Although the sample is not completely representative of the southwestern cities, the data are important especially in light of energy-policy implications. Mail surveys are invariably biased toward higher-income whites. Households at higher socio-economic levels consume the most energy, too, and are the main target population of most conservation policies.

The five cities surveyed had all experienced substantial increases in utility rates in the past few years. The average costs of a 500-kilowatt-hour monthly bill for 1971, 1974, and 1975 in the states and cities sampled are shown in Table 1.3. Albuquerque and Austin have experienced the most substantial rate increases—49 percent and 35 percent, respectively—from 1974 to 1975. This rise is even more substantial when viewed over the period 1971-75, during which a 500-kilowatt-hour bill rose 63 percent in Albuquerque and 74 percent in Austin. Arizona's actual utility costs are higher than those in Texas or New

TABLE 1.3

Average 500-kwh Monthly Electric Bill, Residential Service, 1971-75

	1971	1974	1975
United States*	$11.13	$14.10	$17.93
Texas*	10.18	11.58	14.09
Austin	11.86	15.26	20.60
El Paso	9.25	10.77	13.38
New Mexico*	10.34	11.08	14.89
Albuquerque	11.28	12.35	18.41
Arizona*	12.05	15.78	18.78
Flagstaff	12.69	17.26	19.14
Prescott	12.69	17.43	19.32

*Cities of 2,500 population or more.
Sources: Federal Power Commission 1974, pp. 3, 4, 75, 112, 114; 1975b, pp. 3, 4, 78, 116, 119.

TABLE 1.4

Natural Gas Rates, January 1976
(cents/mcf)

City	Before Fuel Cost Adjustment	After FCA
Austin	2.59	3.07
El Paso	1.17	1.54
Albuquerque	.78	1.11
Flagstaff	1.45	1.45
Prescott	n.a.	n.a.

n.a. = not available.
Source: Southern Union Gas Co. 1976.

Mexico, but the rate of increase has been somewhat slower. Increasing production costs of natural gas have been passed to the consumer in the form of "fuel cost adjustments." Again, Albuquerque and Austin have experienced the largest price increases, as shown in Table 1.4.

Analysis of Southwestern Consumers: Objectives

The historical sketch of energy production, consumption, price, and conservation possibilities clearly shows that the energy future of the United States is in large part dependent on consumer practices, and that residential energy use is not a negligible part of the picture. The present analysis attempts to supply information on the attitudes and beliefs of residential consumers relevant to the energy problem, and to conservation in particular. The book consists of six chapters plus appendix materials. Chapter 2 presents a review of other research on consumer attitudes toward the energy crisis. An annotated bibliography—Appendix B—gives the method, scope of work, and significant findings of recent surveys.

Chapter 3 concerns the findings of the present research that are related to basic attitudes toward energy issues. Beliefs in the energy problem, its nature, duration, and cause are examined. The analysis includes breakdowns by age, sex, race, income, and education. In addition, categories of consumers who differ in energy beliefs are further examined in terms of six social-psychological variables. This approach addresses the question of whether energy-related beliefs are part of a more long-standing belief system.

Chapter 4 analyzes several types of reported behavior. The first issue addressed concerns basic energy information behavior among the subjects. The chapter examines sources of energy information and the frequency with which the repondents discuss energy issues, and to whom they have complained regarding energy problems. Twenty-six items relating to energy conservation are then analyzed, with the focus being to identify those individuals who score high and those who score low on conservation behavior.

In Chapter 5, consumer reactions to various incentive systems are analyzed. The topics discussed are consumer reaction to increases in the price of energy, the time consumers feel is appropriate to recover investments in energy-efficient equipment, and the impact of low-interest government loans in securing energy-efficient equipment. In Chapter 6 some of the policy implications of this research are discussed. All too frequently policy is developed without basic knowledge of the attitudes of the constituents for whom it is formulated. This book, and particularly this chapter, is directed toward bridging the gap between citizen and state in a crucial area: energy consumption and conservation.

The present chapter has reviewed some parameters of the U.S. energy situation in terms of production, consumption, and price. The sample data from five southwestern cities and major demographic data on those cities have been presented. The next step, to be taken in Chapter 2, is a review of what past surveys have found concerning consumer attitudes toward the energy problem.

REFERENCES

Council of Economic Advisors. 1976a. *Economic Report of the President.* Washington, D.C.: U.S. Government Printing Office.

——. 1976b. *International Economic Report of the President.* Washington, D.C.: U.S. Government Printing Office.

Federal Energy Administration. 1976a. *Monthly Energy Review* (September).

——. 1976b. *National Energy Outlook.* Washington, D.C.: U.S. Government Printing Office.

Federal Power Commission. 1974. *Typical Electrical Bills.* Washington, D.C.: U.S. Government Printing Office.

——. 1975a. *All-Electric Homes in the U.S.* Washington, D.C.: U.S. Government Printing Office.

——. 1975b. *Typical Electric Bills.* Washington, D.C.: U.S. Government Printing Office.

——. 1976a. *FPC News* 9 (July 30).

——. 1976b. Opinion and Order, no. 770A. Docket no. RM 7514, November 5.

Ford Foundation. 1974. *Exploring Energy Choices*. Washington, D.C.: the Foundation.

Grebler, Leo; Joan W. Moore; and Ralph Guzman. 1970. *The Mexican American People, the Nation's Second Largest Minority*. New York: The Free Press.

Reed, Raymond D. 1974. *The Impact of and Potential for Energy Conservation Practices in Residential and Commercial Buildings in Texas*. College Station: Texas A&M University.

Science Policy Research Division, Library of Congress. 1975. *Energy Facts*. II. Washington, D.C.: U.S. House of Representatives, Subcommittee on Energy.

Southern Union Gas Company. 1976. Personal contact.

TCA Educational Fund. 1976. *Consumer Watch* 1: (May): 9.

U.S. Bureau of the Census. 1970. *Characteristics of the Population*. Washington, D.C.: U.S. Department of Commerce.

——. 1972. *City and County Data Book*. Washington, D.C.: U.S. Department of Commerce.

U.S. Bureau of Mines. 1975. *United States Energy Through the Year 2000*. Washington, D.C.: U.S. Department of the Interior.

——. 1976. *Mineral Industry Surveys, Weekly Coal Reports*.

2

ENERGY ATTITUDE SURVEYS:
SUMMARY AND EVALUATION

Energy is a vital, societal commodity. Supplies of energy are necessary to the functioning of human groups and form the basis for societal development.

A human population cannot survive without a steady, daily input of energy; and every social and cultural complexity over and above the members' bare survival requires an additional input. An increase in the energy flowing into a system will result in a corresponding increase in goods and services, a larger population, or both. And these, in turn, will stimulate further developments in the group's social structure, its ideology, the variable aspects of its population, and its language (Lenski and Lenski 1974, p. 80).

Man's development, in fact, appears to be coterminous with his effective use of energy. Leslie White (1949, p. 368) has expressed this thesis in the equation $E \times T \overset{a}{=} C$, where C is the degree of cultural development in human societies, E is the amount of energy used, and T is the efficiency of the tools or technology used in the expenditures of that energy.

Despite the critical role that energy plays in society, in recent decades it has not been viewed as a worrisome commodity by the American people. In the public eye, energy was neither in short supply nor expensive. The "out of sight, out of mind" syndrome was abruptly altered, however, in the winter of 1973, when U.S. citizens found themselves waiting in line at gas stations and paying home heating bills that were escalating rapidly. Energy matters were of more than individual concern, too, as U.S. society as a whole faced up to its vulnerable position in terms of dependence on foreign oil imports.

Following the Arab oil embargo, government decision makers and administrators immediately placed emphasis on a dual policy approach to increase domestic energy supplies and to decrease domestic consumption. At this point, however, it became clear that little or nothing was known about consumer attitudes toward the energy crisis; its causes, consequences, and expected duration; or energy consumption in general. Thus, little basis was available for designing effective conservation programs or for estimating their success. In the period between 1973 and 1976, a number of attitudinal surveys were conducted in efforts to obtain a better understanding of the nature of the American consumer, his preferences, and his tolerances. Thus far, however, little time has been spent in scrutinizing these data and comparing results.

This chapter represents an effort to summarize the significant findings of major energy attitude surveys of U.S. citizens. An annotated bibliography (Appendix B) describes the particulars on methods and samples for each survey. The results of this review will provide a basis for comparison of the Southwest attitudinal survey analyzed in the following chapters.

The surveys reviewed share, for the most part, one important trait: they were designed as data-gathering tools to furnish descriptions of consumer attitudes and behavior. Few of the studies were conducted in order to test a hypothesis or hypotheses developed from past social science theory and research. Further, since the surveys were taken during roughly the same time period, little comparison of results and cumulation of data have occurred. As a result, this area of research consists of a wide range of descriptive, and sometimes divergent, findings without a conceptual framework within which these findings can be understood. Without strong conceptual themes to mark the path, the most appropriate way to review and summarize past findings is simply to look for both common and divergent trends in the data. The remainder of this chapter attempts such an evaluation.

SURVEY RESULTS: WHAT ARE THE FINDINGS?

Some of the earliest survey data come from the National Opinion Research Center (NORC) at the University of Chicago. As reported by James Murray et al. (1974), these 1973-74 data show that U.S. consumers were aware of the energy shortage but that few believed it was a major or long-lasting problem. Several other themes appear to have stood the test of later surveys:

1. A majority of consumers held the federal government and large oil companies responsible for the energy situation and, in addition, believed the energy problem to be in large part "contrived," as opposed to "real."

2. Few significant relationships were found between energy attitudes, conservation behavior, and such demographic variables as education, income, and region of residence.

3. Most consumers reported some life-style effects stemming from the energy crisis, but few experienced major changes.

These themes were buttressed by the Opinion Research Corporation opinion polls (ORC 1974-75). Major emphasis in the ORC reports is placed on the fact that within the population few differences on energy attitudes and behavior were found. Rich and poor, young and old appeared skeptical about the energy problem and seemed not to be severely affected by it. On the basis of these early NORC and ORC data, differences in conservation behavior appeared to be price motivated. A basic policy direction was emerging: Since most consumers seem to think and behave similarly with regard to energy matters, economic forces generated in a free-market situation will hold demand to an acceptable level.

Ted Bartell (1974) reports some differences on energy issues by sex and race, but makes no attempt to explain them. The main message is again that energy consumption is best explained in economic terms. In this study, for instance, the only significant relationship with conservation efforts is the reported fear that one's own employment would be negatively affected by the energy problem. Public acceptance of conservation incentives leaned overwhelmingly toward those policies that would require the least personal cost and change in life style.

Robert Perlman and Roland Warren (1975a, b) find more similarities than differences in conservation behavior and attitudes across three metropolitan areas in Connecticut, Alabama, and Oregon. More than 50 percent of their 1974 sample believed that the energy crisis was contrived to boost oil and gas company profits. Again, beliefs were not related to conservation, while price was. All income groups cited price as the major reason for conservation. Reported conservation was greatest where energy rates were highest.

The data described by W. Wayne Talarzyk and Glenn Omura (1975) are similar: consumers blamed oil companies for the energy problem, and such beliefs were not correlated with demographic characteristics. Conservation was generally viewed positively, but no major efforts were reported (despite the fact that more than 33 percent of the sample said that the energy crisis greatly reduced their income). Conservation activities were directly related to income and education. Early reports on the East Urbana study (Hyland et al. 1975) show that rising costs forced some response to the energy crisis, but that conservation efforts made were those involving little change in consumer life style.

Price and Income

As noted, the main motivating mechanism behind conservation efforts appears to be price. Reliance on price to control demand, however, raises questions concerning equity and social justice. Individuals with higher incomes consume more energy and can better afford to keep doing so at higher costs. At the same time, they are better able, if necessary, to cut consumption without significant effects on life style. Nolan Walker and E. Linn Draper (1975) find that from 1972 to 1974, upper-income households in Austin, Texas, increased consumption and appeared likely to continue consuming regardless of price. Middle-income families decreased consumption, while among lower-income households, those who increased consumption were offset by those who decreased, resulting in little net change. The conclusions are similar to those in other studies: higher-income families would not change their energy consumption habits; lower-income families could not do so, since their use was already minimal; and middle-income families would bear the brunt of conservation programs.

Evidence on the nature of the relationship between income, energy beliefs, and conservation effort is diverse, however. Rovend Kilkeary (1975) argues from her survey data that income is the strongest predictor of both energy knowledge and conservation, with the relationship curvilinear. That is, highest knowledge and conservation scores came from the middle-income group. Donald Warren (1974) also finds middle-income households more likely to believe in the reality of the energy crisis and to reduce consumption. Perlman and Warren (1975b), Eunice Grier (1976), and Talarzyk and Omura (1975) report more conservation in higher-income groups. Bonnie Morrison and Peter Gladhart (1976) find family income to be the single best predictor of residential energy consumption. Housing factors directly related to income—such as size, single versus multi-family arrangement, and amount of exterior glass—also related strongly to consumption. Similar factors, however, are not related to belief in the reality of the energy crisis; and belief, in itself, is not found to decrease consumption.

Phillis Thompson and John MacTavish (1976), on the other hand, find three types of consumers differentiated in their sample. The largest group of respondents (more than 50 percent of the sample) is cynical about the nature of the energy problem and the reality of diminishing supply. The behavior of this group is consistent with beliefs—that is, little or no conservation is practiced. This group is made up largely of older persons at lower occupational and educational levels who rely on television for most of their information. A smaller group (20 percent) does believe in real and persistent energy shortages, and has responded to this belief by adopting a variety of conservation measures. Most members of this group are under age 45; work at skilled or professional occupational levels; have college or graduate degrees; and use newspapers, national magazines, and research reports as information sources. The third group is consistently unsure, not knowing who or what to believe or what actions to take. Another survey (Doner 1975) reports that while price remains the most important motivator for conservation measures, those who believe there is an energy crisis behave in a more conservation-minded fashion than those who are cynical and distrustful of the reports of such a crisis.

These varied findings indicate that price and income are indeed important factors in conservation behavior, but that they are not the only ones. These variables work in conjunction with others, and a better understanding of the phenomenon in question will have to take into account interaction effects that may be quite complex. Stages of the family life cycle, for instance, affect energy consumption. Peter Gladhart (1976) reports that families with no children and families where the wife is at least 60 use about 13 percent less energy than families raising children. Those in family-forming age groups, according to Albert Gollin et al. (1976), also report higher levels of concern over the amount of energy consumed in their homes, most probably as a result of financial strain.

Education and Information

Two other important variables are education and social setting. Educational level appears to be positively related to high levels of energy information and to belief in an energy problem (Curtin 1975; Talarzyk and Omura 1975; Zuiches 1975). Mary Stearns (1975) finds that better-educated households view the energy problem as more important, but report less severe personal experiences. On these variables, as on all others, there are divergent findings—such as those by Murray et al. (1974) that report no relationship between education and energy beliefs or conservation practices.

While they do not discuss education directly, David Gottlieb and Marc Matre (1976) report that energy knowledge and belief in an energy problem are directly related to socioeconomic status (measured in terms of occupation, income, and education). They also find that belief in an energy problem is related to conservation of gasoline but not of utilities. David Barnaby and Richard Reizenstein (1975) find that the energy-conscious consumer (an individual who reported willingness to use less home heat) was better-educated and made greater use of media and personal information sources. The issue of education then leads to more specific concerns: knowledge of energy matters and use of information sources.

The last several years have seen a proliferation of energy news and propaganda within the mass media, especially television, ranging from special broadcasts during the peak of the energy crisis to public relations commercials by energy-related companies, as well as a number of public-service broadcasts aimed at educating the public. There is evidence, however, that although educational material and conservation pleas from the media might play a role in forming attitudes and opinions, they have had little effect on consumption behavior. Most consumers reported efforts to cut energy use but, as noted previously, the reason given for conservation was almost always rising costs. While there has been a trend toward more widespread belief in the reality and severity of an energy crisis (Gottlieb and Matre 1976; ORC 1975), there has been little change in the public's claimed efforts to conserve since the energy problem came to the fore (ORC 1975). Speculation on reasons for this phenomenon have important implications for conservation policies.

Two lines of questioning come to mind in considering the effects, or lack of effects, of mass media campaigns to distribute energy information. The first concerns the perceived reliability of media sources. Early survey data show that the American consumer was prone to place blame for the energy crisis on the federal government, particularly Congress, and major energy companies (such as oil and gas). One could predict, then, that consumers would put little faith in energy information generated by those entities. Gottlieb and Matre (1976) found that oil, gas, and electric companies, followed by the government, ranked lowest as accurate and honest information sources. Table 2.1 shows the changes in belief that a source of information was accurate and honest during the period 1974-75. Television and other news media, with the exception of national

TABLE 2.1

Perceptions of Reliability of Information Sources, 1974-75 (percent)

| | Accurate and Honest | | |
| | --- | --- | --- |
	None of the Time	Part of the Time	Most of the Time
Oil companies	44-32	45-55	11-12
Natural gas companies	40-24	49-61	10-15
Electric companies	39-22	51-64	10-14
Government	45-22	44-66	12-13
National newspapers	34-15	43-55	23-29
News magazines	18-8	50-62	32-29
Local newspapers	7-8	53-58	40-34
Radio	7-8	54-66	40-26
Television	4-4	40-52	57-43

Note: Percent changes do not always add to zero because of rounding.
Source: Gottlieb and Matre 1976, pp. 226-34.

newspapers, declined as perceived reliable sources of information. Greater trust of national newspapers, the government, and energy companies was evidenced, although the level of trust remained low. If representative, however, these changes could mean that such energy information strategies as government-sponsored conservation appeals will become increasingly effective.

The ORC data (1974-75) indicate that business ranked low as a reliable information source, with consumer groups, the federal government, and news media ranking high. Interestingly, awareness of FEA dropped slightly over time, despite widespread FEA-sponsored energy advertising. In 1975, 66 percent of the ORC sample were not aware of that agency's existence.

The second major line of questioning has to do with the efficacy of mass media information. Kenneth Novic and Peter Sandman (1974), in a study of how the mass media affect attitudes toward solution of environmental problems, find that high media users consider themselves less informed, view issues as less serious, and prefer less personal solutions to problems than do those who rely more on books, educational courses, and interpersonal communication. The authors' discussion of how this situation came to be is relevant to the energy information situation. By substituting vicarious experience for genuine participation, they argue, the mass media encourage passivity and uninvolvement at the same time that they are informing their audience. A frequent result is an attitude system essentially independent of behavior, much as we have seen in the energy situation—a high level of belief in shortages, but only superficial efforts to reduce consumption. Novic and Sandman (1974, p. 448) explain:

> "I should do this" is a different sort of attitude from "this should be done." The individual who believes that "this should be done" about an important issue—where "this" is something the individual *cannot* do—has an attitude that can never find expression or reinforcement in behavior. However "concerned" that individual may be, the concern is passive.

The mass media tend to stress institutions, rather than individuals, as principal actors in society, thus encouraging these passive attitudes. A frequent theme is that major social problems can be solved only by massive social change. Accurate or not, this viewpoint can be alienating, since it leaves little place for meaningful individual participation. The possibility should be explored that these factors help to account for the lack of individual effort despite high levels of mass media information.

Anthony Downs (1972), in another paper concerning the mass media and environmental issues, provides a second approach relevant to the energy issue. He describes a systematic "issue-attention cycle" regarding public attitudes and behavior concerning most key domestic problems. In this cycle each problem becomes suddenly prominent and then gradually fades from public attention, even though largely unresolved. The cycle begins with the "preproblem stage," in which the condition exists but has not captured the attention of the general public—even though experts or special interest groups are alarmed, and conditions may be worse at this time than they are when the public does become interested. The second stage is one of "alarmed discovery and euphoric enthusiasm," when dramatic events call the existence and evils of a particular problem to the sudden attention of the public. This stage is characterized by general optimism that society will be able to solve or alleviate the problem in a relatively short time. Such an attitude is in line with the American tradition of viewing problems as external to the structure of society, along with faith in science, technology, or the political structure to effect a solution.

The third stage is one of gradual realization of the cost of problem resolution—realization that someone might benefit from the problem and that someone else would have to pay if changes were made. In relation to the energy problem, we have seen questions arise concerning "excess profits" for oil and gas companies, for instance. This stage leads into the fourth, one of gradual decline in public interest. As people realize the difficulty of the situation, they feel discouraged, threatened, or bored; and usually another problem emerges to seize their attention.

The fifth and final "post-problem stage" is one of less attention, with periodic recurrences of interest. New institutions, programs, and policies brought to life in earlier stages ususaly remain and continue functioning; and public interest, while low, is still usually higher than in the preproblem stage (Downs 1972).

The media's role in these stages is one of interaction with the public. As profit-making businesses, the media must provide what the public is interested in, and will drop subjects that bore their audiences. At the same time, mass media have the power to launch a problem from the first stage to the second and to hold it there, at least for a while.

The energy situation, if traced through these stages, presently lies at about the third stage. Consumers seem to realize that costs will be involved in resolving the energy dilemma, but are still hoping a solution will be found before they are forced to pay excessively. There is also indication of a gradual decline in interest and increasing discouragement, which might account for the lack of behavioral change.

The possibility exists, of course, that the energy crisis will not move on through the final stages and be forgotten. In order for problems to go through this "issue-attention cycle," they must meet three criteria: the majority of the population must not be suffering from the problem as much as some minority is; the sufferings caused by the problem must be generated by social arrangements that provide significant benefits to a majority or a powerful minority; and the problem must have no intrinsically exciting qualities to intensify interest frequently. If fuel prices continue to rise so that a majority of the population is affected, or if problem "excitement" is continually reinforced through blackouts, embargoes, and shortages, the cycle may not run its course.

The role of the media in our present energy problem raises many questions. Does media exposure stimulate or satiate interest? What characteristics of the mass media lead to less personal involvement and interest? Can these barriers be overcome to achieve an effective public energy-education program? These and a number of other questions should be answered if further policies involving use of the mass media are to be implemented.

Several experimental studies have been conducted regarding the relationship of energy information to conservation behavior. Thomas Heberlein (1975) found that after a year of steady information, energy consumption in a set of apartment buildings studies was not affected. Another experiment, along slightly different lines, explored the impact of consumption feedback on conservation behavior in a planned urban development of identical dwelling units (Seligman and Darley 1976). In this case, members of the group given daily feedback on their electricity consumption and a score relating this consumption to predicted consumption used 10.3 percent less electricity during the study period than did the control group. Within the feedback group, however, the treatment appeared to be more effective among those participants whose initial consumption was moderate than among high consumers of electricity. A similar experiment (Seaver and Patterson 1976), however, found feedback of consumption rate to have little effect on reducing the use of heating fuel, unless that feedback was coupled with social commendation.

Additional work is needed in the field of mass communication before the reasons for such findings will be well understood. Jane Hass et al. (1975), for example, conducted a small-scale experiment to test the effects of two components of communication messages upon intentions to reduce energy consumption. The authors report that the magnitude of potential personal threat conveyed by a message was more conducive to changed attitudes than was the reported probability of occurrence. Thus, no matter how likely something might be, if it is not expected to have personal consequences, little attitudinal/ behavioral change will occur. Conversely, even a very unlikely event may precipitate change if the perceived consequences are severe. Such findings have direct implication for designing the content of energy-related messages.

Social Settings

Since energy information reaches and influences different people in different ways, it is evident that the social setting in which individuals interact may be an important intervening factor. Donald Warren and David Clifford (1975) show that there are major differences in conservation behavior by type of neighborhood. These researchers classified neighborhoods into six types, from most to least cohesive and active: integral, parochial, diffuse, stepping-stone, transitory, and anomic. Significant differences in reported conservation behavior were found, with the integral neighborhood highest and the anomic neighborhood lowest in conservation effort. The role of neighborhood was often greater than individual income effects. Sources of information used varied by neighborhood type, with integral setting conducive to the use of all information sources from mass media to interpersonal discussion. In anomic settings, at the other extreme, little use was made of any information resource. Warren and Clifford suggest that "Without the 'intervening' role of the social setting which transmits or fails to transmit important shared norms or attitudes, individuals may sow little or no correlation between attitudes and behavior." A major policy-relevant conclusion would be that, in general, conservation strategies will be more successful the more they derive from the local setting.

Opinions over Time

Changes in attitudes and behavior are best examined through longitudinal data, that is, data gathered at two or more points in time. Major studies of this type are only now getting under way. Several analyses, however, do include data over the course of a year or more, and it is on these findings that we now focus.

Murray et al. (1974) used NORC data collected weekly on a national basis since April 1973. They report that, as of 1974, consumers were aware of the energy shortage, although they did not view it as extremely important. A majority of respondents, however, reported some changes in their way of life because of energy problems. Over the course of the year, persons who had difficulties with fuel supplies began to expect serious future problems and more frequently reported negative life changes. Murray et al. suggest "that the evaluations [of importance of the problem] do not become articulated with behavioral events (exposure) until the duration of the situation (shortages) and the pervasiveness of the events have reached a certain threshold" (1974, p. 260). Assignment of responsibility did not vary significantly over time, with the national government and oil and gas companies continuing to receive most blame.

Barnaby and Reizenstein (1975) collected data in February and October 1974. They report an increasing awareness of the insufficiency of energy resources and greater agreement with such policies as resource rationing and obligatory regulation of home temperatures. The ORC polls began in September 1974 and were taken for the next 20 months (ORC 1974-75). Again there is a slight increase in belief that the energy problem is serious. Blame of oil companies for the energy crisis diminishes as blame of the wasteful consumers increases. There is little difference in reported effects, although experience with shortages, limited at first to gasoline, spreads to other fuels. Reported conservation efforts increase, as does the importance of price as the motivation.

David Gottlieb (1974) and Gottlieb and Matre (1976) collected data in 1974, with a follow-up study in 1975. Their data show shifts toward definite belief in an energy crisis over the study year (28 percent to 36 percent). This trend is pronounced for young and middle-aged persons, males, and medium-to-high socioeconomic groups. Also evident is increased agreement over time that the U.S. political structure is capable of "helping Americans through an energy crisis" (45 percent to 52 percent). This change reflects primarily the attitudes of middle-aged-to-older respondents and middle-to-high socioeconomic groups. Other categories of respondents show higher agreement and disagreement, with proportionately fewer "don't know" answers. That is, opinions polarize as individuals begin increasingly to take definite positions in energy matters.

THE AMERICAN CONSUMER: A COMPOSITE PICTURE

The surveys reviewed here are diverse in terms of samples, regions, methods, and findings. Each in itself contributes only a little to an understanding of consumer attitudes or behavior. Nevertheless, in a certain sense, together they accomplish what was intended: large amounts of descriptive data are available on the early years of the energy problem. Much remains to be done in terms of analyzing the data; this review only begins the task.

Despite some contradictory findings on specific variables, a common picture of U.S. consumers in the 1973-75 period emerges. In general, consumers were skeptical when the energy crisis burst upon them. Their impulse was to blame oil and gas companies and the national government for allowing the problem to develop, if not causing it in order to boost fuel prices. These feelings are not surprising, and may have been heightened by the concurrent Watergate scandal. As happens in emergency situations, consumers were willing to accept, on a short-term basis, such measures (including rationing) as were necessary to get through the crisis, whatever its cause may have been. Their skepticism, however, led to disbelief in the information being distributed by energy companies and the government, much of which was inconsistent.

Prices rose and consumption declined. Consumers attributed their changes in energy use to costs, and in large part they were right. Neither the time nor the milieu was conducive to the development of changed attitudes, of a "conservation ethic." As energy use declined, life style was affected severely. They were not sure, for instance, that "everyone" would make efforts to conserve; and they feared the loss of status involved in driving a smaller car, in doing without the energy-driven accoutrements of a middle-class life style. In large part, consumers did not know how to conserve energy efficiently. They did not know such basic facts, for instance, as how much energy was used to heat water in the home, or that they had an option in selecting the temperature of hot water.

Naturally, if they were already well-educated, consumers were better able to turn energy information to their advantage. They also had more access to information. Educated consumers began to report more conservation efforts and more belief in an energy problem. If they had high incomes, however, they might not have bothered to conserve, even though their attitudes were changing. Education and income interacted in their effects on behavior. As a result, certain patterns emerged. Younger consumers had more education and were more prone to believe in the energy problem. On the other hand, they frequently had little reason to conserve, either because they already used little energy or because their life styles were not strained. Perhaps obviously, gasoline was the fuel that most concerned the highly mobile younger consumer.

The segment of the consuming public reporting the most changes in energy-use activities was the middle-income, middle-educated family with children to raise. It is here that education and income may interact in a manner conducive to conservation, as illustrated by Figure 2.1.

Over time, however, our "average" consumers became less skeptical about both the existence of an energy problem and its causes. They began to reflect on the excessive energy used in their homes as they became more knowledgeable about energy matters. At the end of the period we are describing, however, they

FIGURE 2.1

Relation of Age, Education, and Income to Conservation

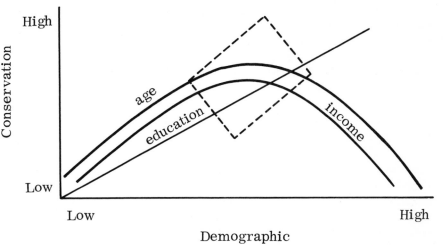

Source: Compiled by the authors.

still were unsure about the duration of the problem; they were mistrustful of government attempts, and lack of attempts, to solve it. They still were wary of the information they received, and they were unwilling to commit themselves to conservation if their neighbors were not going to conserve.

From this picture a number of policy-relevant suggestions emerge. Clearly, the lack of a coherent national policy and program does nothing to encourage individual conservation. Mass media campaigns have been only slightly successful, and should be reexamined in terms of both content and mode of distribution. The greater the amount of information presented on the local level, the more efficacious it will be. It is likely that the same information will be differently received if it is presented personally to the individual rather than through mass media. Of course, time and money set limits here but, for example, groups of speakers and/or prepared programs could be made available to local garden clubs, Rotary clubs, P.T.A. meetings, and so forth for little more money than carefully filmed advertisements distributed to national television networks.

The questions raised here with regard to the energy problem are not unique. They are common to all discussions of social change in a mass society. Much of the theory and data in sociology, psychology, mass communications, and community planning can be brought to bear. And the data collected on energy consumers can be used to strengthen and expand knowledge in those fields.

THE NEXT SURVEY: POINTS FOR COMPARISON

The findings from previous surveys present several points to be considered in the following analysis of southwestern consumers.

1. Beliefs about the energy problem do not appear to differ a great deal across such major demographic groupings as age, sex, race, and income.

2. The only fairly consistent relationship to belief in an energy problem is a positive one with education.

3. Energy beliefs do not appear to be closely associated with energy conservation behavior.

4. The major influence on energy conservation behavior is price, especially for low-to-middle income groups.

Such generalizations can hardly be viewed as encouraging, since they offer little by way of alternative incentive possibilities for conservation. If price is the only mechanism leading to lower consumption, then higher prices it must be, with social inequities handled separately. In the following chapter we will reexamine the nature and distribution of energy attitudes within a major sample of the American public. The focus of this analysis will be on suggesting possible trends in opinion formation.

REFERENCES

Barnaby, David J., and Richard C. Reizenstein. 1975. "Profiling the Energy Consumer: A Discriminant Analysis Approach." Paper presented at ORSA/TIMS Conference, Chicago, Illinois (April).

Bartell, Ted. 1974. "The Effects of the Energy Crisis on Attitudes and Life Styles of Los Angeles Residents." Paper presented at 69th Annual Meeting of the American Sociological Association, Montreal (August).

Curtin, Richard T. 1975. "Consumer Adaptation to Energy Shortages." Ann Arbor: University of Michigan, Survey Research Center. Unpublished.

Doner, W.B., Inc., and Market Opinion Research. 1975. *Consumer Study—Energy Crisis Attitudes and Awareness*. Lansing: Michigan Department of Commerce.

Downs, Anthony. 1972 "Up and Down with Ecology—the 'Issue-Attention Cycle.'" *Public Interest* 28 (Summer): 38-50.

Gladhart, Peter Michael. 1976. *Energy Conservation and Lifestyles:An Integrative Approach to Family Decision-making*. Occasional Paper no. 6, Family Energy Project. East Lansing: Michigan State University, College of Human Ecology, Institute for Family and Child Study.

Gollin, Albert E.; Shirley J. Smith; and Jo Anne S. Youtie. 1976. *New Hampshire Energy Usage Patterns and Consumer Orientations: A Comparative Assessment.* Washington, D.C.: Bureau of Social Science Research, Inc.

Gottlieb, David. 1974. *Sociological Dimensions of the Energy Crisis.* Austin, Texas: Governor's Energy Advisory Council.

Gottlieb, David, and Marc Matre. 1976. *Sociological Dimensions of the Energy Crisis: A Follow-up Study.* Houston: University of Houston, Energy Institute.

Grier, Eunice S. 1976. "Changing Patterns of Energy Consumption and Costs in U.S. Households." Paper presented at Allied Social Science Associations Meeting, Atlantic City, New Jersey (September).

Hass, Jane W.; Gerrold S. Bagley; and Ronald W. Rogers. 1975. "Coping with the Energy Crisis: Effects of Fear Appeals upon Attitudes Toward Energy Consumption." *Journal of Applied Psychology* 60: 754-56.

Heberlein, Thomas A. 1975. "Conservation Information: The Energy Crisis and Electricity Consumption in an Apartment Complex." *Energy Systems and Policy* 1:105-17.

Hyland, Stanley E.; Judith E. Liebman; Demitri B. Shimkin; Richard C. Roistacher; James J. Stukel; and John J. Desmond. 1975. *The East Urbana Energy Study, 1972-1974: Instrument Development, Methodological Assessment, and Base Data.* Champaign: University of Illinois, College of Engineering.

Kilkeary, Rovena. 1975. "The Energy Crisis and Decision-Making in the Family." Springfield, Va.: National Technical Information Service, U.S. Department of Commerce, PB238783.

Lenski, Gerhard, and Jean Lenski. 1974. *Human Societies.* New York: McGraw-Hill.

Morrison, Bonnie Maas, and Peter M. Gladhart. 1976. "Energy and Families: The Crisis and the Response." *Journal of Home Economics* 68 (January): 15-18.

Murray, James; Michael J. Minor; Norman M. Bradburn; Robert F. Cotterman; Martin Frankel; and Alan E. Pisarski. 1974. "Evolution of Public Response to the Energy Crisis." *Science* 184: 257-63.

Novic, Kenneth, and Peter M. Sandman. 1974. "How Use of Mass Media Affects Views on Solutions to Environmental Problems." *Journalism Quarterly* 51, 3:448-52.

Opinion Research Corporation. 1974a. *General Public Attitudes and Behavior Toward Energy Saving.* Highlight Report, I. Princeton, New Jersey: Opinion Research Corporation.

———. 1974b. *General Public Attitudes and Behavior Toward Energy Saving.* Highlight Report, II. Princeton, New Jersey: Opinion Research Corporation.

———. 1974c. *Attitudes and Behavior of Residents in All-Electric Homes.* Highlight Report, III. Princeton, New Jersey: Opinion Research Corporation.

——. 1974d. *Energy Consumption and Attitudes of the Poor and Elderly.* Highlight Report, IV. Princeton, New Jersey: Opinion Research Corporation.

——. 1974e. *Trends in Energy Consumption and Attitudes Toward the Energy Shortage.* Highlight Report, V. Princeton, New Jersey: Opinion Research Corporation.

——. 1975a. *Consumer Attitudes Toward Gasoline Prices, Shortages, and Their Relationships to Inflation.* Highlight Report, VI. Princeton, New Jersey: Opinion Research Corporation.

——. 1975b. *Consumer Attitudes and Behavior Resulting from Issues Surrounding the Energy Shortage.* Highlight Report, VII. Princeton, New Jersey: Opinion Research Corporation.

——. 1975c. *Consumer Behavior and Attitudes Toward Energy-Related Issues.* Highlight Report, VIII. Princeton, New Jersey: Opinion Research Corporation.

——. 1975d. *General Public Attitudes and Behavior Regarding Energy Saving.* Highlight Report, IX. Princeton, New Jersey: Opinion Research Corporation.

——. 1975e. *General Public Attitudes and Behavior Regarding Energy Saving.* Highlight Report, X. Princeton, New Jersey: Opinion Research Corporation.

——. 1975f. *The Public's Attitudes Toward and Knowledge of Energy-Related Issues.* Highlight Report, XI. Princeton, New Jersey: Opinion Research Corporation.

——. 1975g. *General Public Behavior and Attitudes Regarding Vacation and Business Travel, Beverage Containers, Reasons for Using Mass Transit.* Highlight Report, XII. Princeton, New Jersey: Opinion Research Corporation.

——. 1975h. *Energy-Related Attitudes and Behavior of the Poor and the Elderly.* Highlight Report, XIII. Princeton, New Jersey: Opinion Research Corporation.

——. 1975i. *Automobile Usage Patterns.* Highlight Report, XIV. Princeton, New Jersey: Opinion Research Corporation.

——. 1975j. *How the Public Views the Nation's Dependence on Oil Imports; a Possible Natural Gas Shortage This Winter; the Overall Need to Save Energy.* Highlight Report, XV. Princeton, New Jersey: Opinion Research Corporation.

——. 1975k. *Public Attitudes and Behavior Regarding Energy Conservation: Detailed Tabulations by U.S. (January).* Princeton, New Jersey: Opinion Research Corporation.

——. 1975l. *Public Attitudes and Behavior Regarding Energy Conservation: Detailed Tabulations by U.S. (August).* Princeton, New Jersey: Opinion Research Corporation.

Perlman, Robert, and Roland L. Warren. 1975a. *Energy-Saving by Households in Three Metropolitan Areas.* Waltham, Massachusetts: Brandeis University, Florence Heller Graduate School for Advanced Studies in Social Welfare.

————. 1975b. *Energy-Saving by Households of Different Incomes in Three Metropolitan Areas*. Waltham, Massachusetts: Brandeis University, Florence Heller Graduate School for Advanced Studies in Social Welfare.

Seaver, W. Burleigh, and Arthur H. Patterson. 1976. "Decreasing Fuel-Oil Consumption Through Feedback and Social Commendation." *Journal of Applied Behavior Analysis* 2:147-52.

Seligman, Clive, and John M. Darley. 1976. "Feedback as a Means of Decreasing Energy Consumption." Paper presented at 71st Annual Meeting of the American Sociological Association, New York (August).

Sterns, Mary D. 1975. *The Social Impacts of the Energy Shortage: Behavioral and Attitude Shifts*. Washington, D.C.: U.S. Department of Transportation.

Talarzyk, W. Wayne, and Glenn S. Omura. 1975. "Consumer Attitudes Toward and Perceptions of the Energy Crisis." *American Marketing Association 1974 Combined Proceedings*, ed. Ronald C. Curhan, pp. 316-22. Ann Arbor, Michigan: Xerox University Microfilms.

Thompson, Phyllis T., and John MacTavish 1976. *Energy Problems: Public Beliefs, Attitudes, and Behaviors*. Allendale, Michigan: Grand Valley State College, Urban and Environmental Studies Institute.

Walker, Nolan E., and E. Linn Draper. 1975. "The Effects of Electricity Price Increases on Residential Usage by Three Economic Groups: A Case Study." In *Texas Nuclear Power Policies* 5. Austin: University of Texas, Center for Energy Studies.

Warren, Donald I. 1974. *Individual and Communicty Effects on Response to the Energy Crisis in Winter, 1974: An Analysis of Survey Findings from Eight Detroit Area Communities*. Ann Arbor: University of Michigan, Institute of Labor and Industrial Relations, Program in Community Effectiveness.

Warren, Donald I., and David L. Clifford. 1975. *Local Neighborhood Social Structure and Response to the Energy Crisis of 1973-74*. Ann Arbor: University of Michigan, Institute of Labor and Industrial Relations.

White, Leslie A. 1949. *The Science of Culture.* New York: Grove Press.

Zuiches, James J. 1975. *Energy and the Family*. East Lansing: Michigan State University, Department of Agricultural Economics.

3

ENERGY ATTITUDES AND
THEIR CORRELATES

This chapter presents an analysis of the attitudes of persons in five south-western cities regarding the energy problem. As the preceding chapter indicated, past surveys have found little that distinguishes consumers by their energy-related attitudes. The present analysis constitutes an attempt to pursue in detail the correlates of such attitudes, in terms of both demographic and socioeconomic variables and of social-psychological belief structures.

The analysis begins with an overview of sample responses to questions concerning belief in an energy problem, its duration, and its causes. The second section examines differences in beliefs in an energy problem by sex, age, income, race/ethnicity, and education. In the final section of the chapter a different approach is taken. Responses to 35 attitudinal questions are factor-analyzed and respondents are divided into two groups based on their score for each of the seven resulting attitudinal dimensions. This approach allows a more sensitive delineation of trends by examining differences between consumers who feel strongly about energy issues.

AN OVERVIEW OF ENERGY ATTITUDES

Consumers in the Southwest readily admit the existence of energy diffi-culties at the national level. More than 42 percent of the sample strongly agreed with the statement "The United States currently has an energy problem," while another 45 percent indicated they agreed, although not strongly. Only 1 percent of the sample stated thay had no opinion on the issue. Nearly 89 percent of the sample agreed that the energy problem will cause major difficulties for the

United States during the next five years, with 75 percent of the sample fore-seeing difficulties during the next 20 years. Other studies have reported wide-spread belief in an energy problem (Thompson and MacTavish 1976; Murray et al. 1974), a belief that seems to be growing over time (Gottlieb and Matre 1976; ORC 1975).

Belief in the energy problem is difficult to square with the fact that energy consumption continues to increase. Electric power use, for instance, jumped nearly 4 percent in the first quarter of 1975 over what it had been one year before. Precisely, then, at the time consumers were agreeing that an energy problem existed, their behavior appears to have exacerbated the problem. A closer look at the nature of energy attitudes may shed additional light on this area of consumer opinion.

Resource Depletion

It can be argued that meaningful belief in an energy problem arises from an awareness of resource depletion; that is, out of the knowledge that energy shortages are based on actual fuel shortages. The southwestern respondents voiced the opinions shown in Table 3.1 when asked whether the United States is running out of specific types of energy. There is no way of knowing, of course, how these opinions relate to the widespread agreement reported on the existence of an energy problem. It may be, however, that consumers see the present energy situation as tied to the depletion of oil and gas, while expecting intensi-fied production of coal resources to alleviate the energy problem. This inter-pretation is buttressed by the fact that slightly more than 50 percent of the sample felt that "the failure of technology to develop new sources of energy is responsible for the energy problem."

Responsibility for the Energy Problem

Another widely publicized aspect of the energy problem concerns the purported contrivance by oil companies and other energy industries. Was the energy crisis, in short, a trumped-up affair designed solely to raise the prices of energy commodities? Relatively few of the consumers questioned (24 percent) said that the energy problem exists only because high prices are charged for energy. In fact, nearly 60 percent of the respondents said that there would be an energy shortage in the United States even if we were willing to pay a high price for energy. Though price per se is not viewed as the only element in the problem, consumers were quite willing to place blame for the problem on energy industries. When asked if certain sectors were responsible for the energy problem, the respondents gave the answers in Table 3.2.

TABLE 3.1

Respondents' Views on Whether U.S. Fuel Supply Is Dwindling

	Strongly Agree (%)	Agree (%)	No Opinion (%)	Disagree (%)	Strongly Disagree (%)	Total (%)	(n)
Natural gas	18.1	42.1	8.8	26.9	4.2	100	2,382
Oil	17.6	42.5	6.3	29.0	4.6	100	2,375
Coal	4.7	16.4	7.8	50.0	21.1	100	2,360

Source: Compiled by the authors.

TABLE 3.2

Respondents' Views on Responsibility for the Energy Problem

	Strongly Agree (%)	Agree (%)	No Opinion (%)	Disagree (%)	Strongly Disagree (%)	Total (%)	(n)
Petroleum companies	11.7	37.2	8.7	34.2	8.2	100	2,371
Natural gas companies	7.3	31.2	10.7	41.2	9.6	100	2,362
Electricity-producing companies	5.5	25.3	10.9	48.3	10.0	100	2,370
Arab oil-exporting countries	6.5	26.9	9.9	46.2	10.5	100	2,379
U.S. Congress	9.5	37.4	9.0	33.5	10.6	100	2,379
President Ford	2.5	9.9	6.8	45.1	35.7	100	2,375
U.S. consumers	13.8	43.7	6.0	27.7	8.8	100	2,368

Source: Compiled by the authors.

Although responsibility is widely assigned, the petroleum companies in particular are blamed for the energy problem. When asked to agree or disagree with the statement "The petroleum companies did not cause the energy problem, but have taken advantage of it to raise prices," 72 percent of the respondents either strongly agreed or agreed. Only 25 percent of the subjects felt that petroleum companies had been unfairly blamed for the energy crisis. Less than 11 percent of the respondents agreed that petroleum companies have not profited by the energy shortage, but have raised their prices only to cover their added cost.

Blame of the large oil companies has been strong, of course, since the 1973 embargo. Most surveys have found that a majority of respondents hold the oil companies responsible for the energy problem (Murray et al. 1974; Bultena 1976). More than 50 percent of Robert Perlman and Roland Warren's sample (1975) agreed that the energy crisis was contrived to boost oil and gas company profits. David Gottlieb (1974) reported that his respondents most frequently listed "desire for profit" by oil and gas companies as the major reason for the shortage of energy. In fact, more respondents believed that the shortage was part of a political scheme than believed the world was running out of energy fuels.

The data in Table 3.2 also show, however, that blame was placed more frequently on the consumer than anywhere else. If this pattern is taken to reflect awareness of "energy excessive" life styles and habits prevalent among U.S. consumers, it can be viewed with hope from the standpoint of energy conservation. Whether or not the present data show a trend or simply an anomaly is something we cannot ascertain. ORC (1975) data show that blame of oil companies has diminished, with an increase in blame of consumer wastefulness. Gottlieb and Matre (1976) report similar indications. It is reasonable to argue that the data presented here show a change of attitudes over time, and that the early "scapegoating" attitude toward oil companies is giving way to somewhat different conceptions of the causes of the energy problem. It should be added, however, that there is no reason to predict that oil companies are going to "get off the hook" in public opinion any time in the foreseeable future.

A hypothetical picture of consumer attitudes is emerging from the data presented thus far. Blame for the energy problem is widespread. Generally, consumers realize that the energy sources now relied on most heavily (oil and gas) are being depleted, but feel that other resources (for example, coal) are available if the technology were developed to fully exploit them. Futhermore, consumers seem to be aware that Americans use a great amount of energy, and that their energy "dependence" is exploited by energy industries that, if not directly responsible for the present problem, have at least turned that problem to an advantage in the marketplace.

WHO BELIEVES IN THE ENERGY PROBLEM?

Having extrapolated this much from the responses of the whole sample of respondents, the question now to be answered is whether there are differences

of energy attitudes within the sample population. Are certain groups of particular ages, education levels, or income levels more likely to believe in an energy
problem than others are? In order to find an answer to this question, one key
attitude question is examined in relation to the major demographic characteristics of sex, age, race, education, and income.

The statement reads, "There will be no energy shortage in the United
States as long as we are willing to pay a high price for energy." It has two
advantages as a dependent variable reflecting energy attitudes. First, in order to
express a belief that the energy problem is not solely price-related, respondents
had to respond negatively (that is, "disagree"). This answer requires somewhat
more consideration than does the more general statement "The United States
currently has an energy problem." The fact that it was easier for the respondent
merely to check off "agree" on the latter statement is indicated by 88 percent
agreement, whereas 60 percent expressed disagreement with the first question.
Second, the chosen statement implies a longer time dimension in the energy
problem than is reflected in the latter item, which focuses on the existence of
an energy problem now. Using the above statement to reflect "belief in an
energy problem," we find some interesting differences in opinion among our
sample respondents.

Opinions by Sex

As evident from the data in Table 3.3, women are slightly more likely than
men to believe in an energy problem. The difference is not particularly significant (gamma = -.136; eta = .003), but it is not without precedent. James
Zuiches (1975) found that women were more likely to believe in the energy
crisis and generally were more energy-aware.

TABLE 3.3

Respondents' Belief in an Energy Problem, by Sex

	Women (%)	Men (%)
Belief in energy problem	64	58
Lack of belief	30	37
No opinion	6	5
Total		
Percent	100	100
n	632	1,651

Source: Compiled by the authors.

Opinions by Age

Belief in an energy problem is decidely less among older groups, although the statistical significance of the differences is, again, lacking (gamma = -.09; eta = .004) (see Table 3.4). Other studies show inconsistent results on the age variable. Gottlieb and Matre (1976) found that respondents who were skeptical of a real energy problem were generally younger. To the contrary, Phyllis Thompson and John MacTavish (1976) found older respondents to be more cynical, while younger respondents more frequently expressed belief in real and persistent shortages. Mary Stearns (1975) reported that older respondents believed that energy shortages would be of shorter duration than did younger respondents.

TABLE 3.4

Respondents' Belief in an Energy Problem, by Age

	Under 30 Years (%)	30-60 Years (%)	Over 60 Years (%)
Belief in energy problem	68	59	55
Lack of belief	26	36	37
No opinion	5	5	8
Total			
Percent	99	100	100
n	321	829	199

Source: Compiled by the authors.

Even though the present data show differences that are not highly significant, the percentages convey the impression that younger people are more accepting of an energy problem. This finding, if indicative of a trend, suggests that belief may become more widespread in the future. The question arises, however, as to the role education plays in the relationship, since younger groups are generally the recipients of more education than older groups are.

Opinions by Education

The relationship between education and belief in an energy problem is even stronger than the age-belief relationship (see Table 3.5). Although the statistical relationship is still not impressive (gamma = -.21; eta = .02), if one is trying to predict trends, it is apparent that education is a positive influence on belief in an energy problem. In most past surveys (such as Zuiches 1975; Thompson and MacTavish 1976; Curtin 1975), belief in an energy problem and education have shown positive correlations.

TABLE 3.5

Respondents' Belief in an Energy Problem, by Education

	No High School Diploma (%)	High School or Trade School Graduate (%)	College (%)	Graduate or Professional Degree (%)
Belief in energy problem	49	45	64	67
Lack of belief	46	48	32	27
No opinion	5	7	4	6
Total				
Percent	100	100	100	100
n	154	499	935	703

Source: Compiled by the authors.

Opinions by Income

Some differences are found in belief by income category (total family income); this would be expected because of the high correlation between income and education. A greater percentage of respondents at higher income levels believe in an energy problem (see Table 3.6).

TABLE 3.6

Respondents' Belief in an Energy Problem, by Income

	Less Than $10,000 (%)	$10,000-$19,999 (%)	$20,000 or More (%)
Belief in energy problem	58	57	65
Lack of belief	36	37	31
No opinion	6	6	4
Total			
Percent	100	100	100
n	572	996	670

Source: Compiled by the authors.

However, this income variable does little to help explain differences in belief (gamma = -.04; eta = .001). Donald Warren (1974) and Rovena Kilkeary (1975) report a curvilinear relationship between income and energy belief, with the middle class most likely to believe in the energy crisis. Other surveys generally report that the relationship is linear, with respondents at higher income levels more likely to believe in the energy problem (Gottlieb and Matre 1976; Perlman and Warren 1975; Thompson and MacTavish 1976; Warren and Clifford 1975).

Opinions by Race/Ethnicity

Also predictable are the differences by race and ethnicity, since the black and Mexican-American groups rank low in education (see Table 3.7).

TABLE 3.7

Respondents' Belief in an Energy Problem, by Race/Ethnicity

	White (%)	Black (%)	Mexican-American (%)
Belief in energy problem	61	52	54
Lack of belief	34	40	41
No opinion	5	8	5
Total			
Percent	100	100	100
n	1,907	52	236

Source: Compiled by the authors.

It must be noted that in all groups more than 50 percent of the respondents agree than an energy problem exists. This is an important point that will be returned to later. The next section, however, will examine the impact of the energy problem on the family.

THE ENERGY PROBLEM AND
EFFECTS ON THE FAMILY

Stearns (1975) reports that high social status is related to a better understanding of the energy problem in the abstract. That is, individuals at higher income and educational levels are more likely to perceive impacts stemming from the problem that do not focus on personal effects. Consumers of lower social status are more prone to see problems through the lens of their own experiences. There is nothing unreasonable about this; lower-to-middle-income people often have more problems created for them by some occurrence, and their opinions are shaped by those problems. In general, however, whatever the social class level, people consistently accept general principles while failing to come to a consensus on specifics (see Prothro and Grigg 1960).

It is important here to better differentiate the general and the specific. Ralph Turner (1954) defines "values" at the general level as those which refer to something regarded as favorable or unfavorable that can be secured or attained. "Values" on the specific level are "norms," which are "prescriptions regarding behavior and belief and prohibitions against certain patterns of behavior and belief" (Turner 1954, p. 301). In any given society there will be greater agreement on values than on norms. It is easier, let us say, to agree that "democracy" is a good thing to pursue than to agree on rules and procedures that give specific meaning to the value "democracy." Similarly, it is possible to find high agreement on "conservation" as a favored goal, but very little consensus on what everyday kinds of rules should be followed to achieve that goal. Clearly, this is what is found with regard to belief in an energy problem. Almost everyone in the sample agrees that the United States has an energy problem. People at upper income and educational levels are more likely to express concern over a longer-term problem that higher prices alone cannot cure. Individuals at lower social class levels are more concerned about what present prices are doing to them and their families.

Sample respondents were asked, "If your electric (gas) bill has gone up, how has that affected your family?" Response categories were "It really had no effect on us"; "We had to make a few adjustments, but our style of life was not affected"; "Our life was less comfortable and convenient, but it was not serious"; and "We had to make serious changes in our daily habits." The data show results that, at first glance, may appear inconsistent (see Tables 3.8 and 3.9).

Those individuals who report being most affected by increasing energy costs are less likely to express belief in the reality of the energy problem. When viewed in light of the preceding discussion of the effects of education and income on belief, however, this apparent contradiction seems more reasonable. Those consumers who report that rising energy prices have had no effect on their family are more likely to be in higher income brackets than those who have been forced, by prices, to make serious changes. Higher-income individuals are more likely to believe in the energy problem. It is also possible that those who are making serious changes in their daily habits see the problem as a purely economic one resulting from oil and gas company contrivance, or other political reasons. The belief question used here, it must be remembered, essentially asks whether willingness to pay a high price could solve the energy problem. Lower-income consumers may have read into this "ability to pay a high price"—and indeed, if these people could pay a high price, they would certainly have less of a problem.

TABLE 3.8

Effects of Electric Bill Increases

	No Effect (%)	Few Adjustments (%)	Less Comfortable and Convenient (%)	Serious Changes (%)
Belief in energy problem	64	62	56	50
Lack of belief	31	32	40	45
No opinion	5	6	4	5
Total				
Percent	100	100	100	100
n	452	1,021	647	157

Note: Gamma = –.123; eta = .008.
Source: Compiled by the authors.

TABLE 3.9

Effects of Natural Gas Bill Increases

	No Effect (%)	Few Adjustments (%)	Less Comfortable and Convenient (%)	Serious Changes (%)
Belief in energy problem	62	61	56	50
Lack of belief	32	33	39	44
No opinion	6	6	5	6
Total				
Percent	100	100	100	100
n	633	957	564	106

Note: Gamma = –.084; eta = .004.
Source: Compiled by the authors.

ATTITUDINAL STRUCTURES AND TRENDS

The relationships of the demographic variables of age, education, income, sex, and race/ethnicity to belief in an energy problem can be appreciated from the previous data, but little of the statistical variation in belief can be explained on the basis of those variables (see Table 3.10 for regression coefficients). There are several possible reasons for this lack of significance. As we shall later argue,

TABLE 3.10

Regression of Demographic Variables on Energy Belief

	Multiple R-Square	R-Square Change
Age	.00810	.00810
Education	.04359	.03549
Sex	.04439	.00080
Income	.04471	.00032

Source: Compiled by the authors.

it is important in itself that a great deal of agreement exists among the American people. For now, however, we should note that, first, there could be methodological problems. The questionnaire items used to tap attitudes simply may not have been successful. Second, and more important, the energy problem is relatively new to U.S. consumers; and it is entirely possible that attitude structures among different segments of the population are still tentative and not closely integrated into the individual's belief structure. As Charles Tittle and Richard Hill (1970, p. 469) explain, "The individual encountering a situation which is characterized by unfamiliar continuencies is not likely to have a well-structured attitudinal organization relevant to behavior in that situation."

It may be more useful in this case—both for explaining and for predicting—to examine individuals who do seem to have strong opinions about energy matters. The ability to ascertain trends is enhanced, in fact, not so much by examining a total population as by analyzing the extreme components of that population (Myers 1972). This is the approach taken here.

Mode of Analysis

Responses to 35 attitudinal statements were factor-analyzed, using an equimax rotation. The resulting six factors, with the component statements and their factor loadings, are given in Table 3.11. One additional factor is discussed in the text, for a total of seven dimensions. Taken together, the factors can explain 52 percent of the statistical variance among the items. Only three items did not load on a factor.

Responses on the statements within each factor were treated in a Likert fashion to give an overall score. The range of scores for each statement is 1 = strongly agree, 2 = agree, 3 = no opinion, 4 = disagree, 5 = strongly disagree. The factor scores can be used to divide respondents into two extreme groups, dropping middle-range subjects. For instance, respondents who strongly agreed

TABLE 3.11

Attitudinal Factors and Item Loadings

Factor	Item	Loading
Extent of the energy problem	The United States is running out of oil	.81156
	The United States is running out of natural gas	.80310
	The United States currently has an energy problem	.64883
	The United States is running out of coal	.55627
	U.S. consumers are responsible for the energy problem	.53399
	There will be no energy shortage in the United States as long as we are willing to pay a high price for energy	−.51569
	Our energy problem exists only because we are being charged high prices for energy	−.50325
Present and long-term impacts of the energy problem	I feel the energy problem will cause major difficulties for the United States during the next five years	.81712
	I feel the energy problem will cause major difficulties for the United States during the next 20 years	.66933
	The energy problem has put a substantial strain on my budget	.52953
Responsibility for the energy problem	The petroleum companies are responsible for the energy problem	.60924
	The natural gas companies are responsible for the energy problem	.59975
	The electricity-producing companies are responsible for the energy problem	.59422
	The state government is responsible for the energy problem	**.57816**
	The Arab oil-exporting countries responsible for energy problem	**.54536**
	The U.S. Congress is responsible for the energy problem	**.53801**
Corporation practices regarding the energy problem	The natural gas companies have been unfairly blamed for the energy problem	.71396
	The electric companies have been unfairly blamed for the energy problem	.71114
	The petroleum companies have been unfairly blamed for the energy problem	.69542
	The electric companies have raised their prices only to cover their added costs and have not profited by the energy shortage	.60688
	The petroleum companies have raised their prices only to cover their added costs and have not profited by the energy shortage	.49342
	The natural gas companies have raised their prices only to cover their added costs and have not profited by the energy shortage	.47634
Who has taken advantage of the energy problem?	The petroleum companies did not cause the energy problem, but have taken advantage of it to raise prices	.77069
	The electric companies did not cause the energy problem, but have taken advantage of it to raise prices	.75256
	The natural gas companies did not cause the energy problem, but have taken advantage of it to raise prices	.74741
Who has tried to solve the energy problem?	Congress has done all it can to solve the energy problem	.73731
	President Ford has done all he can to solve the energy problem	.70251
	The natural gas companies have done all they could to solve the energy problem	.69343
	The petroleum companies have done all they could to solve the energy problem	.58541
	The electric companies have done all they could to solve the energy problem	.45417

Source: Compiled by the authors.

with all items in a factor would have an average score of 1. The factor responses are not distributed normally, but are skewed in one direction or the other. On one set of questions, for instance, most respondents may have disagreed, skewing the distribution to the right. On another set of items, responses may have been skewed to the left in agreement. Each factor distribution of scores was examined for what seemed to be the most logical breaking point, in order to establish two sets of respondents: one in high agreement on a factor and the other in high disagreement. While the exact number of people examined varies by factor because of the different degrees of skewness, the polar groups roughly represent the top third and the bottom third of the respondents, with the middle group omitted from the analysis. The last group represents those people who more frequently had no opinion on an item and less frequently gave a "strongly" agreeing or disagreeing response.

In the following sections these polar groups of consumers are examined in terms of demographic characteristics and of certain underlying belief patterns. The questionnaire contained several items that make up standardized scales measuring social-psychological attributes and political attitudes. Six scales were included. First, ten items compose a dogmatism scale, which measures "the extent to which the person can receive, evaluate, and act on relevant information received from the outside on its own intrinsic merits, unencumbered by irrelevant factors in the situation arising from within the person or from the outside" (Rokeach 1960, p. 57). The more dogmatic an individual, the more he interprets information in terms of his existing beliefs and the less that belief structure is open to new information. V.C. Troldahl and F.A. Powell's (1965) short version of Rokeach's original scale was used.

Second, a scale developed by Herbert McClosky (1958) was included as a measure of general, not political, conservatism. High scores on this scale are correlated with the acceptance of conventional social attitudes and emphasis on "duty, conformity, and discipline" (Robinson, Rusk, and Head 1972, p. 94). High scorers tend to be poorly educated.

Two political alienation scales developed by Marvin E. Olsen (1969) represent the third and fourth scales used here. One is intended to measure attitudes of incapability and the other, attitudes of discontentment. Incapability encompasses feelings of guidelessness and powerlessness. Within his social system, the individual feels that he can do little because of the constraints imposed on him. Discontent involves feelings of dissatisfaction and disillusionment—"voluntarily chosen by the individual as an attitude toward the system" (Robinson and Shaver 1969, p. 181).

The fifth and sixth scales represent attitudes toward government and toward large private companies, as measured by two instruments devised at ORC (1960). The six-item big-government scale covers intervention or participation in certain activities, such as controlling corporate profits or guaranteeing farm prices. The big-business scale, also of six items, measures attitudes toward the concentration of power in the private sphere, the role of large companies in the nation's growth, and profits.

Each of the attitudinal factors is discussed below. Differences between the groups of consumers who felt strongly one way or the other on each dimension are identified and summarized in text tables. Complete data are provided in Appendix C.

Factor 1—Extent of the Energy Problem

As can be seen from Table 3.11, this attitudinal factor is scored on the basis of seven statements. Respondents who strongly agreed with all of the items (the last two are reverse-scored) can be considered to believe that the energy problem is a major one for the country. These people not only believe the country to be running out of energy resources, but also hold U.S. consumers responsible and do not think that the problem can be solved by paying higher prices. On the other hand, respondents who disagreed with the statements are the "nonbelievers," those who do not agree that the nation has a significant energy problem. The subjects in the first group compose roughly 34 percent of the sample. Their average score (see "Mode of Analysis" for an explanation of the scores) for the factor was 1.85, with a range of 1 to 2.14. The second group—the "nonbelievers," makes up 29 percent of the sample, with an average score for the factor of 3.57 and a range of 3.14 to 5.

Table 3.12 shows the major differences between the two groups of consumers to be consistent with the earlier analysis of belief in an energy problem. The "believers" are characterized to a greater extent than are the "nonbelievers" by high income, low age, and high education. For instance, 67 percent of consumers with a yearly family income of $25,000 or more agreed that energy constitutes a major problem for the country. This compares with about 50 percent of the consumers in the lower income groups. In terms of age, more than 70 percent of the under-30 group and 58 percent of the 30-39 group rated the energy problem as significant. This compares with 46 percent of the 40-49 and 50-60 groups, and 48 percent of those respondents over 60.

There were very dramatic differences in the attitudes of the four education groups. Less than 30 percent of the persons in the two least educated groups, those with less than high school education and those who went to trade school, felt that the nation really has a serious energy problem, while 58 percent of the subjects in the graduate school-educated category felt that the nation really does have a serious energy problem.

The race/ethnic status of the subjects also clearly differentiates the two groups of subjects. Fifty-seven percent of the white respondents felt that the nation has a real energy problem, while 64 percent of the blacks and 62 percent of the Mexican-Americans felt that the nation's energy problem is not very significant. Tables C.1-C.5 present the complete demographic data for factor 1.

TABLE 3.12

Factor 1—Extent of the Energy Problem

	Major Problem	Minor Problem	Significance Level
Demographic variables			
Family income	High income	Lower-middle and middle incomes	.000[a]
Age	Younger	Middle age and older	.000[a]
Education	Higher education	Lower education	.000[a]
Sex	No difference	No difference	n.s.[a]
Race/ethnicity	White	Black and Mexican-American	.000[b]
Social-psychological variables			
Dogmatism	Less dogmatic	More dogmatic	.000[b]
Conservatism	Less conservative	More conservative	.000[b]
Political incapability	Less feeling of incapability	More feeling of incapability	.000[b]
Discontent with politics	Less discontent	More discontent	.000[b]
Big business	More positive	More negative	.000[b]
Big government	More negative	More positive	.004[b]

[a]Chi-square test.
[b]F-test.
n.s. = not statistically significant.
Source: Compiled by the authors.

The social-psychological scales reflect the social class standing of those respondents more strongly agreeing that an energy problem exists (see Table C.6 for complete data). Consumers who have more "agree" scores on this factor tend to be less dogmatic, less conservative, less politically alienated, more favorable toward big business, and less favorable toward big government. These social-psychological attributes are corollaries of high income and educational levels.

Factor 2—Present and Long-Term Impacts of the Energy Problem

As indicated in Table 3.11, factor 2 is made up of three items dealing with whether the energy problem has had a substantial impact on the subject's budget and whether energy-related problems will exist in the United States during the next five years and the next 20 years. The group of respondents who feel that the United States has, and will continue to have, major energy difficulties accounts for 39 percent of the sample. They have a mean response score for the three items of 1.33 and a range of 1 to 1.66. Those subjects who feel that the problem is more transitory in nature have a score range of 2.66 to 5, with a mean of 3.22. (See "Mode of Analysis" for an explanation of these scores.) This group composes 23 percent of the sample. Table 3.13 summarizes the findings with respect to factor 2, and shows that three of the five demographic variables and four of the six social-psychological variables significantly differentiated the subjects who felt the energy problem was more of a major problem for themselves and the country from the group who felt this was not true.

TABLE 3.13

Factor 2–Present and Long-Term Impacts of the Energy Problem

	Substantial	Not Substantial	Significance Level
Demographic variables			
Family income	Middle income	Lowest, highest income	.01[a]
Age	Younger	Older	.000[a]
Education	No difference	No difference	n.s.[a]
Sex	Slightly more female	Slightly more male	.03[a]
Race/ethnicity	No difference	No difference	n.s.[a]
Social-psychological variables			
Dogmatism	No difference	No difference	n.s.[b]
Conservatism	No difference	No difference	n.s.[b]
Political incapability	More feeling of incapability	Less feeling of incapability	.000[b]
Discontent with politics	More discontent	Less discontent	.000[b]
Big business	More negative	More positive	.000[b]
Big government	More positive	More negative	.000[b]

[a]Chi-square test.
[b]F-test.
n.s. = not statistically significant.
Source: Compiled by the authors.

More than 50 percent of each income group felt that the energy problem would be long-standing. This attitude was more pronounced, however, among the three middle-income segments. More than 65 percent of the $5,000-$19,999 groups felt strongly in that direction, compared with 41 percent of the lowest income group and 47 percent of the highest income category (data for this factor are given in Tables C.7-C.12). This finding runs counter to what might have been expected on the basis of factor 1, where middle-income respondents were less likely to report that the energy problem was of major significance. It is possible to speculate on several differences among the items in factors 1 and 2 that might underlie the shift in response. The key item here is probably the third one in factor 2, "The energy problem has put a substantial strain on my budget." Middle-income consumers are more apt to agree with that statement.

As was the case for factor 1, the younger the respondents are, the more likely they are to believe that the energy problem presents major difficuties for themselves and for the country. The percentage of these respondents goes from nearly 73 percent among those under 30 to 48 percent among those over 60.

Although not statistically significant, it is worth noting that education, too, works in the same direction on the first two factors, with respondents at higher educational levels more committed to the fact of energy problems. On factor 2, however, there is less of a spread among the education categories. For example, 55 percent of those respondents with less than a high school education responded on factor 2 that the energy problem was major, compared with 22 percent saying on factor 1 that the problem was significant. Fewer respondents at the other end (attended graduate school) agreed with factor 2 than they did with factor 1 (62 percent versus 72 percent). This pattern is, again, probably related to the "strain on budget" item in factor 2, and to the fact that more middle-income respondents (who generally have less education) responded strongly to that statement.

Females showed more energy concern on this factor than did males, but the differences are small. In terms of race/ethnicity, no differences are found. This reflects the fact that a much higher percentage of blacks and Mexican-Americans showed concern on factor 2 than on factor 1; for example, 66 percent compared with 36 percent in the case of blacks. This shift is predictable, given the earlier discussion of income and education. Different types of people are expressing strong feelings on factor 2 than did on factor 1; and this, we have argued, is in large part due to the fact that personal consequences of the energy problem are reflected in the second factor.

The shift in respondents is clearly seen, too, when we examine social-psychological correlates of energy attitudes. Whereas on factor 1, the social-psychological scores reflect the large percentages of higher social class respondents who felt the energy problem to be significant, on factor 2 the scores reflect the greater tendency on the part of the middle classes to feel strongly about the energy problem. In the present case, more political alienation, and

more hostility toward big business and less toward big government, are characteristic of respondents who agreed that the energy problem will last and has
impacts on their budgets.

Factor 3—Responsibility for the Energy Problem

The third attitudinal factor is made up of six statements that deal with
responsibility for the energy problem (Table 3.11). This factor does not focus
on who is to blame but, rather, on how willing the respondent is to assign
responsibility. The six statements concern whether petroleum, natural gas, and
electricity-producing companies, state government, Arab oil-producing countries,
and the U.S. Congress are responsible for the energy problem. The sample is
divided, for present purposes, into a group that is highly willing to assign responsibility (33 percent of the total) and a group that is reluctant to assign
responsibility (32 percent). The middle 33 percent of the sample is omitted,
as usual. The "heavy blamers" have an average score (see "Mode of Analysis")
of 2.16, with a range of 1 to 2.66, and the "low blamers" have a mean score of
4 and a range of 3.66 to 5 on the factor.

As can be seen from Table 3.14 (complete data are in Tables C.13-C.18),
respondents from lower social classes generally are more willing to assign responsibility. In terms of income, for instance, 59 percent of the lowest income group
were "heavy blamers," compared with 33 percent of the highest income segment.
In terms of education, nearly 68 percent of the respondents with less than a
high school education fall in the "willing to assign responsibility" group, whereas
only 34 percent of respondents who had attended graduate school are in that
group. Minority groups (blacks, Mexican-Americans, women) are more heavily
represented here. The social-psychological scores also reflect this social class
distribution. The "heavy blamers" are more politically alienated (incapability
and discontent), more negative toward big business and positive toward big
government, more dogmatic in their beliefs, and generally more conservative.

The younger respondents are the most represented of all age groups in
the "heavy blame" category, but the differences are not as large as they are
among income or education levels. Fifty-four percent of the respondents under
30, compared with 42 percent of the consumers over 60, scored high on willingness to place responsibility.

Another factor was identified that also concerns responsibility for the
energy problem. This factor was made up of the statements to the effect that
President Ford/President Nixon was responsible for the problem. That factor
is not treated at length here because only 12 percent of the sample held Ford
responsible and 20 percent thought Nixon was responsible. The data show,
however, that among those respondents who hold the presidents responsible, we
find the same representation of low-income, low-education, younger respondents
that we have seen on the general responsibility factor.

TABLE 3.14

Factor 3—Responsibility for the Energy Problem

	Willing to Assign Responsibility	Not Willing to Assign Responsibility	Significance Level
Demographic variables			
Family income	Lower income	Higher income	.000[a]
Age	Younger	Older	.03[a]
Education	Lower education	Higher education	.000[a]
Sex	Slightly more females	Slightly more males	.05[a]
Race/ethnicity	More black and Mexican-American	More white	.000[a]
Social-psychological variables			
Dogmatism	More dogmatic	Less dogmatic	.000[b]
Conservatism	More conservative	Less conservative	.000[b]
Political incapability	More feeling of incapability	Less feeling of incapability	.000[b]
Discontent with politics	More discontent	Less discontent	.000[b]
Big business	More negative	More positive	.000[b]
Big government	More positive	More negative	.000[b]

[a]Chi-square test.
[b]F-test.
Source: Compiled by the authors.

Factor 4—Corporation Practices

This factor is based on six statements dealing with whether the natural gas, electricity, and petroleum companies have been "unfairly blamed for the energy crisis," and whether these companies "have raised their prices only to cover their added costs and have not profited by the energy shortage." Unlike the previous factor which allowed an assessment of the respondent's willingness to assign blame, the present factor permits an evaluation of the respondent's attitudes toward a particular sector: energy industries. "Strongly agree" responses to the items composing this factor can be taken to represent positive opinions about corporation practices in the energy field, and "strongly disagree" responses can be assumed to reflect negative opinions. The scores on factor 4 are skewed in favor of negative opinions, with very few positively inclined consumers. For this reason, although the sample was divided so that roughly

33 percent of the respondents are in each of the two extreme groups, the
"positive" group contains considerable "don't know" answers. These answers
can be taken to reflect at least a feeling of neutrality. The lowest scorers (35
percent of the sample) represent, then, consumers who are either positively
inclined toward energy industries or neutral about them. The mean score for this
group is 2.81, with a range of 1 to 3.16. The highest scorers (36 percent of the
sample) are those consumers who are very negative about the role that energy
industries have played in the present energy problem. These respondents have
a mean score or 4.67, and a range of 4 to 5. (See "Mode of Analysis" for
explanation of scores.)

The data (see Tables C.19-C.24) reveal rather clear differences between
individuals who strongly feel that private corporations are in some measure to
blame for the energy problem and those who are not negative about such businesses. As the results summarized in Table 3.15 indicate, consumers who are

TABLE 3.15

Factor 4—Corporation Practices

	Positive-Neutral Attitude	Negative Attitude	Significance Level
Demographic variables			
Family income	Extreme lower and extreme upper income levels	Middle income	.000[a]
Age	Older	Younger	.000[a]
Education	Highest and lowest education levels	Middle education levels	.000[a]
Sex	No difference	No difference	n.s.[a]
Race/ethnicity	No difference	No difference	n.a.[a]
Social-psychological variables			
Dogmatism	No difference	No difference	n.s.[b]
Conservatism	More conservative	Less conservative	.02[b]
Political incapability	Less feeling of incapability	More feeling of incapability	.000[b]
Discontent with politics	Less discontent	More discontent	.000[b]
Big business	More positive	More negative	.000[b]
Big government	More negative	More positive	.000[b]

[a]Chi-square test.
[b]F-test.
n.s. = not statistically significant.
Source: Compiled by the authors.

particularly negative toward big business are middle-income, younger, and have an average education. In the middle-income categories ($5,000-$19,999), nearly 60 percent of the respondents hold negative attitudes toward oil, gas, and electric companies. At the two income extremes, more than 50 percent of the respondents are positive or neutral. The same pattern is evidenced in terms of education level, with roughly 60 percent of the people in the two middle education levels apparently hostile toward energy industries. Again, more than 50 percent of both the lowest and highest education groups were positive or neutral.

This pattern is explicable because of the manner in which the sample was divided for this particular factor. The idea behind this part of the analysis is to compare the extremes—to analyze consumers who feel very strongly in one way about an issue against consumers who feel very strongly in the other way about that issue. Respondents who were noncommittal are eliminated from the analysis. In the present case, however, there were not enough respondents who felt strongly positive toward energy industries for the comparison. Included with them, for the sake of numbers, then, are respondents with neutral feelings— that is, who responded "don't know" to the statements composing factor 4. In every survey, however, it is the respondent with the lowest income and education who is most likely to respond "don't know." Therefore, the positive-neutral group is composed of two distinctly different types of people: on the one hand, there are the well-educated, older respondents who have favorable opinions about private industry; on the other hand, there are the low-educated, poor respondents who are unable to express an opinion about corporate practices and the energy problem.

The social-psychological scale scores reflect these differing proportions of respondents. The presence of the older, high-income respondents on the positive-neutral side is evidenced in the lower feelings of political alienation, the more positive attitudes toward private corporations, and the less positive stance on big government.

Factor 5—Who Has Taken Advantage of the Energy Problem?

This factor is related to the previous one (regarding attitudes about energy industries), and is made up of three items that determine whether respondents feel that pertroleum, natural gas, and electric companies have taken advantage of the energy problem to raise the price of their products. Responses on this factor are again skewed, with the bulk of the sample agreeing that energy industries have turned the energy crisis to their advantage. Fifty-one percent of the sample constitutes the group that has rather strong "agree" feelings. The mean factor score for these consumers is 1.72, with a range of 1 to 2. The comparison group contains only 20 percent of the sample, with a mean score of 4 and a range of 3.3 to 5. These latter respondents disagree that energy industries have taken advantage of the energy problem to raise prices.

TABLE 3.16

Factor 5—Who Has Taken Advantage of the Problem?

	Energy Companies Have Taken Advantage of the Problem	Energy Companies Have Not Taken Advantage of the Problem	Significance Level
Demographic variables			
Family income	No difference	No difference	n.s.[a]
Age	No difference	No difference	n.s.[a]
Education	No difference	No difference	n.s.[a]
Sex	No difference	No difference	n.s.[a]
Race/ethnicity	No difference	No difference	n.s.[a]
Social-psychological variables			
Dogmatism	No difference	No difference	n.s.[b]
Conservatism	No difference	No difference	n.s.[b]
Political incapability	More feeling of incapability	Less feeling of incapability	.007[b]
Discontent with politics	More discontent	Less discontent	.000[b]
Big business	More negative	More positive	.000[b]
Big government	More positive	More negative	.000[b]

[a]Chi-square test.
[b]F-test.
n.s. = not statistically significant.
Source: Compiled by the authors.

As Table 3.16 indicates, differences between the two groups are not statistically significant (see Tables C.25-C.30 for full data). The data show, however, the kinds of trends that would be expected on the basis of earlier factors. In general, those respondents who feel the energy companies have not taken advantage tend to be higher-income, higher-education, older, and white. Lower social class, young to middle-age, and minority respondents are more heavily represented in the group with negative opinions about energy companies. For instance, 77 percent of those making less than $5,000 feel the companies have taken advantage, compared with 67 percent of those making more than $25,000. Eighty-one percent of respondents with less than a high school education tend to blame the companies, compared with 70 percent of those people with graduate school education. As with other factors, the table shows the political alienation and attitudes toward big business and government characteristic of lower-income and less-educated people. Respondents who feel that energy companies have taken advantage of the nation's energy problem to raise prices are more likely to feel incapable of dealing with the political system and are more discontent with it. These consumers are more negative toward large companies in general, and are more likely to feel that the government should take an active role in regulating the private sector.

There is an aspect of the present findings that we have not yet presented, and that is particularly relevant here. Thus far the focus in this analysis has been on finding differences in the population in terms of energy attitudes. The other side of the coin is equally important: a substantial homogeneity of opinion is a critical social factor. When one finds, as we have here, that large majorities of all income, all education, all age, and all minority groups agree on an issue, the inescapable conclusion is that a major societal problem or issue exists. When the population is in accord that energy companies have taken advantage of the energy problem to raise prices, then social policy eventually must reflect that accord. These findings are much more important to public and private decision makers than when only a particular group of citizens feels strongly about an issue.

Factor 6—Attempts to Solve the Energy Problem

The final factor to be discussed concerns whether attempts have been made to solve the energy problems. Five items are included, which concern whether Congress, President Ford, petroleum, natural gas, and electric companies "have done all they could to solve" the problem. Obviously, respondents who strongly agreed with all five items are expressing the attitude that the energy problem is being addressed by public and private sectors. As might be expected, the sample is extremely biased away from that attitude. The group that feels strongly that the above entities are not doing all they can to solve the problem constitutes 35 percent of the sample, with a mean factor score of 4.7 and a range of 4.4 to 5. The other end of the distribution is not very extreme. We have taken the lower 25 percent of the same sample for purposes of comparison. This group has a mean score of 3.11 and a range of 1 to 3.6. Clearly, this group is "neutral," characterized by large numbers of "don't know" responses. It cannot be said that these consumers feel strongly that public and private efforts are being made to solve our energy problem.

Differences between the two groups are striking (see Table 3.17 and data in Tables C.31-C.36). Respondents who feel that not enough is being done about the energy situation are of the upper social class and are younger. The other group is characterized as low-income, low-education, and older. As pointed out previously, this is due in large part to the tendency among lower socio-economic status people to give "don't know" responses. Particularly interesting, however, is the age distribution. Whereas 74 percent of respondents under 30 feel strongly that substantial efforts to solve the energy problem have not been made, that percentage decreases across age groups to a low of 29 percent in the age group over 60. This pattern is consistent with earlier findings. Young people are both more believing of an energy problem and more discontent about what is being done about it.

TABLE 3.17

Factor 6—Attempts to Solve the Energy Problem

	Attempt Has Been Made/or Neutral	No Attempt	Significance Level
Demographic variables			
Family income	Lower income	Middle and upper income	.001[a]
Age	Older	Younger	.000[a]
Education	Lower education	Higher education	.000[a]
Sex	No difference	No difference	n.s.[a]
Race/ethnicity	No difference	No difference	n.s.[a]
Social-psychological variables			
Dogmatism	More dogmatic	Less dogmatic	.000[b]
Conservatism	More conservative	Less conservative	.000[b]
Political incapability	No difference	No difference	n.s.[a]
Discontent with politics	Less discontent	More discontent	.001[b]
Big business	More negative	More positive	.000[b]
Big government	More negative	More positive	.05[b]

[a]Chi-square test.
[b]F-test.
n.s. = not statistically significant.
Source: Compiled by the authors.

As usual, the low income and low education among respondents who feel that attempts to solve the problem have been made are associated with more political alienation and negative feelings toward big business. The high percentages of older people in this category raise the scores on dogmatism, conservatism, and positive attitudes toward big business.

ENERGY USE AS A SOCIAL ISSUE

The preceding analysis shows that different people view various energy issues in different ways. Younger consumers are, across the board, concerned about the energy problem. People with high income and education are aware of the difficult energy situation, and are less likely to blame energy companies or to feel that such companies have taken selfish advantage of the nation's problem. Respondents in the middle and lower social classes are less perceptive of such factors as resource depletion, but are very aware of the energy problem when it is couched in personal terms (such as budget difficulties due to higher energy costs). Low-income and low-education consumers are more uncertain about the role energy industries have played in the energy crisis and about efforts to solve the nation's energy difficulties.

As we mentioned previously, however, the import of these findings pales against the significance of the finding that a great deal of homogeneity among the American people seems to exist on energy attitudes. The majority of subjects —regardless of how much money they make, how much education they have, or how old they are—believe that the country has an energy problem of some lasting significance and that not enough is being done by public or private sectors to solve it. No clearer mandate could be given to the policy maker.

REFERENCES

Bultena, Gordon L. 1976. *Public Response to the Energy Crisis: A Study of Citizens' Attitudes and Adaptive Behaviors.* Ames: Iowa State University.

Curtin, Richard T. 1975. "Consumer Adaptation to Energy Shortages." Ann Arbor: University of Michigan, Survey Research Center. Unpublished.

Gottlieb, David. 1974. *Sociological Dimensions of the Energy Crisis.* Austin, Texas: Governor's Energy Advisory Council.

Gottlieb, David, and Marc Matre. 1976. *Sociological Dimensions of the Energy Crisis: A Follow-up Study.* Houston: University of Houston, Energy Institute.

Kilkeary, Rovena. 1975. "The Energy Crisis and Decision-Making in the Family." Paper prepared for the National Science Foundation. Washington, D.C.: National Technical Information Service.

McClosky, Herbert. 1958. "Conservatism and Personality." *American Political Science Review* 52: 27-45.

Murray, James; Michael J. Minor; Norman M. Bradburn; Robert F. Cotterman; Martin Frankel; and Alan E. Pisarski. 1974. "Evolution of Public Response to the Energy Crisis." *Science* 184: 257-63.

Myers, Jerome J. 1972. *Fundamentals of Experimental Design*. Boston: Allyn and Bacon, pp. 162-65.

Olsen, Marvin E. 1969. "Two Categories of Political Alienation." *Social Forces* 47: 288-99.

Opinion Research Corporation. 1975. *Public Attitudes Toward Energy*. Highlight Report series. Princeton, New Jersey: Opinion Research Corporation.

Perlman, Robert, and Roland L. Warren. 1975. *Energy-Saving by Households of Different Incomes in Three Metropolitcan Areas*. Waltham, Massachusetts: Brandeis University, Florence Heller Graduate School for Advanced Studies in Social Welfare.

Prothro, James W., and Charles M. Grigg. 1960. "Fundamental Principles of Democracy: Bases of Agreement and Disagreement." *Journal of Politics* 22: 276-94.

Robinson, John P., and Phillip R. Shaver. 1969. *Measures of Social Psychological Attitudes*. Ann Arbor: University of Michigan, Institute for Social Research.

Robinson, John P.; Jerrold G. Rusk; and Kendra B. Head. 1972. *Measures of Political Attitudes*. Ann Arbor: University of Michigan, Institute for Social Research.

Rokeach, M. 1960. *The Open and Closed Mind*. New York: Basic Books.

Stearns, Mary D. 1975. *The Social Impacts of the Energy Shortage: Behavioral and Attitude Shifts*. Washington, D.C.: U.S. Department of Transportation.

Thompson, Phyllis T., and John MacTavish. 1976. *Energy Problems: Public Beliefs, Attitudes, and Behaviors*. Allendale, Michigan: Grand Valley State College, Urban and Environmental Studies Institute.

Tittle, Charles R., and Richard J. Hill. 1970. "Attitude Measurement and Prediction of Behavior: An Evaluation of Conditions and Measurement Techniques." In *Attitude Measurement*, ed. Gene F. Summers, pp. 468-78. Chicago: Rand McNally.

Troldahl, V.C., and F.A. Powell. 1965. "A Short Form Dogmatism Scale For Use in Field Studies." *Social Forces* 44: 211-14.

Turner, Ralph. 1954. "Value-Conflict in Social Disorganization." *Sociology and Social Research* 38: 301-08.

Warren, Donald I. 1974. *Individual and Community Effects on Response to the Energy Crisis in Winter, 1974*. Ann Arbor: University of Michigan, Institute of Labor and Industrial Relations.

Warren, Donald I., and David T. Clifford. 1975. *Local Neighborhood Social Structure and Response to the Energy Crisis of 1973-74*. Ann Arbor: University of Michigan, Institute of Labor and Industrial Relations.

Zuiches, James J. 1975. *Energy and the Family*. East Lansing: Michigan State University, Department of Agricultural Economics.

4

ENERGY-RELATED
BEHAVIOR

In Chapter 3 the energy attitudes of the respondents were examined. In this chapter we move to a discussion of behavior as it relates to the energy problem. Several specific areas are analyzed, including communication and information patterns and conservation efforts.

COMMUNICATION AND INFORMATION
PATTERNS

Several parts of the original questionnaire were devoted to energy information and communication variables. First, the subjects were asked to state where they obtain most of their information on energy matters. Second, they were questioned as to how frequently they discuss the energy problem with a number of specified individuals. Third, the subjects were asked how often they complain about energy problems to public and private officials. After a preliminary examination of these factors (below), the subjects are classified into "complainers" and "noncomplainers." These two groups are then compared across the attitudinal and demographic variables introduced in the last chapter.

Sources of Information

The respondents were asked to check their first, second, and third most important sources of information on energy matters. As the data in Table 4.1 indicate, newspapers were the first most important source of information for over 40 percent of the respondents. Television was second, and news magazines third (13 percent). As a second choice, television was first (36 percent), newspapers second (30 percent), and news magazines third (15 percent). For third choice, radio and news magazines both received more than 20 percent of the responses, while television and newspapers had almost 16 percent each.

TABLE 4.1

First, Second, and Third Most Important Sources of Information on Energy Matters

Source	First Choice	Second Choice	Third Choice
Newspapers	42.6	30.2	15.7
Television	29.3	36.1	15.9
High school courses	.2	.6	2.4
Radio	3.6	10.1	20.2
News magazines	13.2	15.0	23.7
College courses	2.4	1.6	3.0
Friends and family	1.2	2.7	10.7
Government literature	1.6	2.1	4.6
Other	5.9	1.6	3.8
Total percent	100.0	100.0	100.0
n	2,256	2,238	2,194

Source: Compiled by the authors.

The press, television, and radio appear to provide most of the energy information that consumers receive. Other sources (such as high school or college courses and government literature) are relatively unimportant sources of information on energy matters for most of the subjects. These findings are not surprising in a "mass media" era. It may be recalled from Chapter 2 that the Gottlieb and Matre (1976) and ORC (1975) studies found that consumers place high reliability on television, radio, and newspapers. While the mass media are successful in reaching consumers, one can question the nature of information relayed. In Chapter 2 it was suggested that mass media messages do not convey a role for individual participation in the resolution of societal problems. The data presented here show that policy makers cannot avoid using the media, since citizens rely on it heavily for information. The mass media are an effective means of distributing information, but the content of that information must be carefully shaped.

Amount of Discussion of Energy Problems

The consumers were asked to state how frequently they had discussed the energy problem with their spouses, children, friends, people at work, and neighbors. The response categories ranged from "very frequently" to "not at all." Table 4.2 shows that the subjects were most likely to talk with their spouses about the energy problem, and then with people at work or friends. Nearly 66 percent of the respondents claimed to talk, either frequently or very frequently, about energy matters with these other people. If their answers are at all valid, then we can conclude that people are interested enough in energy matters to discuss them often. Interestingly, neither children nor neighbors are frequently engaged in conversation on energy topics.

TABLE 4.2

How Often the Subjects Discussed the Energy Problem with Others

Discussed with	Very Frequently (%)	Frequently (%)	Occasionally (%)	Not at All (%)	Totals (%)	(n)
Spouse	33.1	36.8	24.1	6.0	100.0	2,021
Children	9.7	22.6	36.9	30.8	100.0	1,840
Friends	18.1	43.5	35.8	2.6	100.0	2,283
People at work	23.4	37.3	30.1	9.2	100.0	2,211
Neighbors	11.3	20.5	47.0	21.2	100.0	2,208

Source: Compiled by the authors.

These data are useful from a policy point of view because they indicate what is and what is not a "meaningful community" for the consumer. It seems that the neighborhood setting is less frequently a locus for social interaction. In the automobile age, people are not constrained by geographic distances to having their next-door neighbors as friends or even conversationalists. The workplace, on the other hand, evidently provides a situation of information exchange. Several fairly simple ideas can be extrapolated for use in planning energy conservation programs. Energy information distributed in the schools probably will not filter "up" to parents very rapidly. On the other hand, "energy awareness" programs in the schools are important for children, particularly if there is little discusssion of the subject at home. The work situation might be effectively used for dissemination of energy materials and as a place for energy-related talks or discussion sessions. With employer cooperation, programs could be developed to show employees conservation techniques applicable to the work environment and how these techniques are transferable to their homes.

Even simple data such as these can be helpful in furthering an understanding of consumer behavior and in promoting programs designed to modify that behavior. Another question of some understandable sensitivity concerns consumers who complain about energy bills and shortages. It would be useful to know more about consumers who actively complain, both in order to know how they might be helped and in order to better understand which citizens are being affected by the energy situation. The following section addresses this issue.

COMPLAINTS ABOUT ENERGY PROBLEMS

Respondents were asked to indicate how often during the last two years they had complained to their congressman, gas company, electric company, petroleum company, newspapers, state officials, or federal officials. The response categories included "I have not complained during the last two years," "I have complained once," "I have complained two or three times," "I have complained four or five times," "I have complained more than five times."

TABLE 4.3

Frequency of Energy-Related Complaints During the Past Two Years

	No Complaints (%)	1 Complaint (%)	2-3 Complaints (%)	4-5 Complaints (%)	> 5 Complaints (%)	Total (%)	Total (n)
Congressman	81.6	9.0	7.1	1.2	1.1	100.0	2,210
Gas company	72.6	12.8	10.8	1.8	2.0	100.0	2,221
Electric company	71.4	12.0	11.3	2.6	2.7	100.0	2,234
Petroleum company	84.1	5.3	5.6	2.1	2.9	100.0	2,146
Newspaper	93.9	3.3	1.4	.9	.5	100.0	2,128
State officials	88.7	5.0	4.1	1.1	1.1	100.0	2,141
Federal officials	90.5	4.0	3.2	1.1	1.2	100.0	2,121

Source: Compiled by the authors.

As the data in Table 4.3 indicate, most of the subjects did not complain very often. The two most frequent recipients of complaints were gas and electric companies. Nearly 30 percent of the subjects stated that they had complained at least once to their electric company. Newspapers and federal and state officials received the fewest complaints.

Complainers vs. Noncomplainers

The responses on the frequency-of-complaint questions were summed in a Likert manner to give an overall score for each consumer. Any subject who had two or more total complaints, regardless of where he complained, was defined as a "complainer," while a respondent with one or no complaints was defined as a "noncomplainer." Thrity-three percent of the sample falls in the "complainer" category. These groups were compared in terms of the attitudinal factors, demographic variables, and social-psychological beliefs discussed in Chapter 3. Table 4.4 presents a summary of the findings concerning the differences between the complainers and noncomplainers.

Energy Attitudes

There is a fairly distinct pattern of energy attitudes among those consumers who complain about energy problems and those who do not. On factor 1, regarding belief about the extent and nature of the energy problem (see Table 3.1 for a description of attitude factors and items), complainers feel less strongly than do noncomplainers. That is, noncomplainers are more likely to agree that the United States is running out of basic energy resources, that the energy problem is not due solely to prices, and that consumers are responsible for the problem. Complainers were more likely to agree, however, on factor 2—that the energy problem will continue to affect the nation in coming years and that the problem has put a strain on their budget. Complainers were, further, more willing to assign responsibility for the energy situation (factor 3).

Subjects classified as complainers are also more negatively inclined toward energy industries (factor 4) and more convinced that oil, gas, and electric companies have taken advantage of the energy problem to raise their prices (factor 5). Finally, complainers are more likely to believe that more could be done to solve the energy problem (factor 6).

Demographic Factors

Given the above attitudinal trends, one would expect to find more lower-income and lower-education respondents among the complainers. None of the demographic variables, except race/ethnicity, significantly differentiates the two groups, however. Roughly 32 percent of all white respondents fell in the complainer group, compared with 45 percent of the black respondents and 43 percent of the Mexican-American consumers.

Social-Psychological Variables

Past research has shown that complaining is directly related to a high sense of political efficacy—which, in turn, is usually a function of high education

TABLE 4.4

Comparison of Complainers and Noncomplainers

	Complainers	Noncomplainers	Significance Level
Attitudinal factors			
Extent of energy problem	Felt less strongly about existence of the energy problem	Felt more strongly about existence of the energy problem	.000[a]
Present and long-term impacts of the energy problem	Felt more strongly about the impact	Felt less strongly about the impact	.000[a]
Responsibility for the energy problem	More willing to assign blame	Less willing to assign blame	.000[a]
Corporation practices	More negative	Less negative	.000[a]
Who has taken advantage of the energy problem	Felt more strongly that companies have taken advantage	Felt less strongly that companies have taken advantage	.05[a]
Who has tried to solve the problem	Felt more strongly that no one had tried to solve	Felt less strongly that no one had tried to solve	.000[a]
Demographic variables			
Family income	No difference	No difference	n.s.[b]
Age	No difference	No difference	n.s.[b]
Education	No difference	No difference	n.s.[b]
Race/ethnicity	More minority	Less minority	.05[b]
Sex	No difference	No difference	n.s.[b]
Social-psychological variables			
Dogmatism	No difference	No difference	n.s.[a]
Conservatism	No difference	No difference	n.s.[a]
Political incapability	No difference	No difference	n.s.[a]
Discontent with politics	More discontent	Less discontent	.000[a]
Big business	More negative	More negative	.000[a]
Big government	More intervention	More intervention	.000[a]

[a] = F-test.
[b] = Chi-square test.
n.s. = not statistically significant.
Source: Compiled by the authors.

(Friedmann 1974). This means that low-education/low-income citizens are generally the least likely to complain. Few differences are found here between complainers and noncomplainers in terms of social-psychological attributes. Somewhat greater feelings of discontent with the political system and more negative feelings toward big business are found among the complainers. Complainers are also more willing to support government intervention in the private sector. These attributes are generally associated—as was seen in Chapter 2— with low income and education. Thus, some indication exists that lower socio-economic level citizens may be taking a more active part in complaining about their energy situation and bills than research in other areas has shown.

We have seen, to this point, that most consumers rely on the mass media for energy information and that a majority of respondents report frequent discussion of the energy problem—quite a few consumers have complained about energy problems. The major behavioral issue is yet to be addressed: how much conservation is there, and who is conserving?

CONSERVATION EFFORTS

The crucial questions in energy policy concern whether or not people are trying to conserve energy and what can be done to influence them to conserve. The analysis that follows addresses the first issue, while Chapter 5, on conservation incentives, involves an examination of the latter. The respondents were asked to indicate their efforts to reduce energy consumption on 26 different statements. The statements ranged from "turned thermostats down in winter (up in summer)" to "replaced appliances with more energy-efficient ones." The response categories were "substantial efforts" = 1. "moderate efforts" = 2, "slight efforts" = 3, "no effort" = 4, and "not applicable" = 5. In this analysis, the "not applicable" responses were eliminated.

On many items the majority of respondents reported that they had made substantial efforts to conserve. Such responses fell off sharply on items that involved more expense or more effect on habits and life style. The responses on conservation measures that are among easiest to perform, in terms of time and money, are shown in Table 4.5. More than 50 percent of the respondents (excluding those who checked "not applicable") said they had made substantial to moderate efforts in all of these areas. On only one statement did more than 20 percent of the consumers reply that they had made no effort to conserve.

Table 4.6 includes items that require more effort on the part of the consumer. In some cases significant extra work may be required, such as hanging out clothes, defrosting a freezer, or washing dishes by hand. Other items mean a lower comfort level for the respondent and his or her family, such as not using an air conditioner or using less hot water. Reported conservation is lower

TABLE 4.5

Consumer Responses on Easy Conservation Measures

Conservation Question	Substantial Efforts (%)	Moderate Efforts (%)	Slight Efforts (%)	No Effort (%)	Total (%)	(n)
Tried to always turn out light when not needed	65.3	28.7	5.2	.8	100.0	2,371
Turned thermostats down in winter	48.5	37.6	10.0	3.9	100.0	2,310
Turned thermostats up in summer	46.1	31.7	11.4	10.8	100.0	1,632
Opened drapes during the day, closed at night during the winter	51.1	26.3	10.9	11.7	100.0	2,262
Closed fireplace damper when not in use	56.0	19.1	9.4	15.5	100.0	1,267
Turned off pilot to furnace in the summer	56.4	7.7	4.5	31.4	100.0	2,088
Turned off decorative yard lights	59.0	14.2	7.7	19.1	100.0	1,074
Washed only full loads in clothes washer	55.7	25.9	10.0	8.4	100.0	2,155
Turned heat off during the day while away in the winter	54.2	22.5	12.3	11.0	100.0	2,179
Turned air conditioning off during the day while away in the summer	66.6	17.6	8.3	7.5	100.0	1,802
Closed off unused rooms	52.2	21.0	12.3	14.5	100.0	1,985

Source: Compiled by the authors.

TABLE 4.6

Consumer Responses on Conservation Measures Requiring More Work/Less Comfort

Conservation Question	Substantial Efforts (%)	Moderate Efforts (%)	Slight Efforts (%)	No Effort (%)	Total (%)	Total (n)
Replaced light bulbs with bulbs of lower wattage	19.2	26.0	19.0	35.8	100.0	2,287
Used dishwasher less	25.0	25.5	19.7	29.8	100.0	1,320
Watched TV less	9.0	19.7	20.9	50.4	100.0	2,183
Turned down water-heater thermostat setting	13.2	16.3	16.0	54.5	100.0	2,181
Used less hot water	13.8	24.0	23.9	38.3	100.0	2,209
Turned dishwasher off before dry cycle	18.1	11.1	11.8	59.0	100.0	1,208
Defrosted freezer more often	11.8	17.6	18.5	52.1	100.0	1,430
Hung clothes to dry rather than using clothes dryer	28.4	11.8	13.1	46.7	100.0	1,949
Used fans and opened windows instead of using air conditioning	39.8	26.2	16.2	17.8	100.0	1,890

Source: Compiled by the authors.

TABLE 4.7

Consumer Responses on Conservation Measures That Require an Expenditure

Conservation Question	Substantial Efforts (%)	Moderate Efforts (%)	Slight Efforts (%)	No Effort (%)	Total (%)	Total (n)
Installed storm windows	10.3	7.9	6.8	75.0	100.0	1,451
Installed Thermopane windows	4.8	3.0	3.6	88.6	100.0	1,446
Purchased insulating drapes	19.4	11.8	7.4	61.4	100.0	1,799
Added insulation to attic	21.2	8.5	6.3	64.0	100.0	1,373
Installed weatherstripping on doors and windows	20.7	20.6	18.0	40.7	100.0	1,852
Replaced appliances with more energy-efficient ones	6.8	10.9	13.8	68.5	100.0	1,174

Source: Compiled by the authors.

here than in Table 4.5, but it is important to note that in no case does the group of consumers making substantial or moderate efforts fall below 25 percent. The highest "no effort" responses are where they could be expected: watching television less (which does not save a great deal of energy anyway), turning down the hot water setting (which consumers may not know how to do, or know how much energy can be saved), and hanging clothes or defrosting freezers (which entail substantial work).

A third set of conservation items receives even fewer reported efforts. These items require people to spend money in making their homes more energy-efficient. (See Table 4.7.) Efforts are most frequently made on improvements that are both most advertised and relatively inexpensive: insulation, weather-stripping, and insulating drapes. The energy-saving benefits of storm windows or Thermopane windows may not yet be evident to consumers in the Southwest. The item concerning replacement of appliances is difficult to interpret because efforts here are constrained by existing investments. Few consumers will replace a functioning appliance with another simply to be more energy-efficient. The important question not answered here is whether consumers are prepared to buy more energy-efficient appliances as replacements when their present ones wear out.

It seems clear that as conservation becomes more time-consuming and expensive, consumers are less committed to it. This is not surprising, as earlier surveys have found (see Chapter 2). More important, though, is the fact that large numbers of consumers believe themselves to be conserving energy in a number of ways. To be addressed now is the question of who these consumers are.

WHO CONSERVES?

The 26 conservation statements in Tables 4.5-4.7 were factor-analyzed using a varimax rotation to determine whether different types of conservation behavior could be discerned. The analysis revealed four factors that, as expected, reflected different levels of time and dollar investment. Table 4.8 shows the items and their factor loadings. The only item that did not load on any of the four dimensions was "used fans and opened windows instead of using air conditioning." The four factors explain 48.3 percent of the variance among the statements.

Once the factors were determined, a respondent's scores for each dimension were summed and divided by the number of items to which he had replied, excluding any items marked "not applicable." Respondents with factor scores of 1-2 (substantial or moderate effort to conserve) were classified as "more conserving," while those with scores of 3-4 (slight or no effort) were classified as "less conserving." In the analysis that follows, these two groups are broken

TABLE 4.8

Conservation Factors and Factor Loadings

Factor	Item	Loading
Adjusted	Turned thermostats up in summer	.83806
thermostat	Turned thermostats down in winter	.81179
and lighting	Tried always to turn out lights when not needed	.58272
Improved use	Turned air conditioning off during the day while away	
patterns	in summer	.69269
	Turned off decorative yard lights	.66393
	Turned heat off during the day while away in winter	.63395
	Turned off pilot to furnace in the summer	.60091
	Washed only full loads in clothes washer	.56491
	Closed off unused rooms	.55598
	Closed fireplace damper when not in use	.47558
	Opened drapes during the day, closed at night	
	during the winter	.44174
Reduced	Used less hot water	.68880
energy-	Watched TV less	.65141
consuming	Used dishwasher less	.62232
activities	Turned down water-heater thermostat setting	.61967
	Defrosted freezer more often	.57713
	Hung clothes to dry rather than using clothes dryer	.55298
	Replaced light bulbs with those of lower wattage	.53243
	Turned dishwasher off before dry cycle	.51764
	Replaced appliances with more energy-efficient ones	.45987
Installed	Added insulation to attic	.68881
energy-	Installed Thermopane windows	.65324
conserving	Installed storm windows	.64049
materials	Purchased insulating drapes	.54452
	Installed weatherstripping on doors and windows	.52842

Source: Compiled by the authors.

down by the demographic variables (income, education, sex, age, and race/ethnicity), the social-psychological variables (see Chapter 3 for description), and by frequency of discussion on energy issues. This approach allows evaluation of how those consumers who conserve more differ from those who conserve less.

Factor 1—Adjusted Thermostat and Lighting

The first factor is made up of three items that are relatively easy to accomplish: turning off unneeded lights and adjusting the thermostat for home heating and cooling. Since a large majority of the sample reported that they did make these types of conserving efforts, the "more conserving" group contains 83.5 percent of all respondents. The "less conserving" group is made up of 16.5 percent of the sample.

Since almost everyone reports conservation efforts, the differences are slight; but reported efforts decrease slightly as income and education increase,

TABLE 4.9

Consumer Comparison on Factor 1, Adjusted Thermostat and Lighting

	More Energy-Conserving	Less Energy-Conserving	Significance Level
Demographic variables			
Income	Majority of all income levels	More higher-income	.004[a]
Age	No difference	No difference	n.s.[a]
Education	No difference	No difference	n.s.[a]
Sex	Majority of both males and females	Slightly more males	.000[a]
Race/ethnicity	No difference	No difference	n.s.[a]
Social-psychological variables			
Dogmatism	No difference	No difference	n.s.[b]
Conservatism	No difference	No difference	n.s.[b]
Political incapability	No difference	No difference	n.s.[b]
Political discontent	No difference	No difference	n.s.[b]
Big business	More negative	More positive	.04[b]
Big government	More intervention	Less intervention	.04[b]
Communication patterns			
Spouse	Very frequently	Frequently	.000[b]
Children	Frequently	Frequently-occasionally	.000[b]
Friends	Frequently	Frequently-occasionally	.000[b]
People at work	Frequently	Slightly less than frequently	.000[b]
Neighbors	Frequently-occasionally	Occasionally	.000[b]
Complainers' index	More complaints	Fewer complaints	.02[b]

[a] = Chi-square test.
[b] = F-test.
n.s. = not statistically significant.
Source: Compiled by the authors.

as Table 4.9 indicates (data on this factor are reported in Tables D.1-D.6). Approximately 88 percent of respondents with family incomes of less than $5,000 are classified as more conserving, decreasing to 79 percent of respondents in the $25,000 or more category. By education, the percent in the more conserving group drops from 89 percent in the lowest education group to 81 percent in the highest group. Age and race/ethnicity show no differences. Slightly fewer men than women rank as high conservers.

No differences are found between the two groups of consumers on the social-psychological scales, except for slightly more negative feelings about big business and more positive sentiments toward big government among the more conserving—a pattern frequently accompanying lower income levels in this analysis. Interestingly, people who report more conservation efforts also report more frequent discussion of energy matters.

Factor 2—Improved Energy-Use Patterns

The second factor includes eight items that require more effort and planning on the part of the consumer, such as washing only full loads of clothes, turning off the heat or air conditioning when the home is not occupied, and closing off unused rooms (see Table 4.8). These conservation efforts, it should be noted, ordinarily will not interrupt the consumer's life style or make large demands on his time. The more conserving group of respondents constitutes nearly 69 percent of the sample—a majority, but substantially less than that on the earlier factor. The less conserving group makes up 31 percent of the respondents.

The summary of findings in Table 4.10 shows patterns similar to those on the first factor, except that they are more pronounced. Lower-income and less-educated consumers are heavily represented in the more conserving group. Nearly 85 percent of those people with incomes under $5,000, compared with 51 percent at the other extreme, are classified as more conserving. Similarly, 80 percent of respondents at the lowest education level have high conservation scores, whereas only 61 percent of the highest education group have high scores. Age appears to be important also: 80 percent of consumers under 30 report substantial-to-moderate conservation efforts, but only 66 percent of the older age groups report similar efforts. The high conservers contain overrepresentations of women, blacks, and, on this factor, Mexican-Americans.

The social-psychological pattern becomes more distinct when viewed with the demographic trend. The more conserving respondents are more dogmatic and more conservative, and have stronger feelings of political incapability and discontent. In addition, the more conserving subjects are, the more mistrustful they are of big business and the more they feel government should be more active in regulating society. Finally, these people are more involved with energy as a social issue, which is reflected in the fact that they frequently discuss energy issues with other people and make more complaints to both public and

TABLE 4.10

Consumer Comparison on Factor 2, Improved Use Patterns

	More Energy-Conserving	Less Energy-Conserving	Significance Level
Demographic variables			
Income	Lower income	Higher income	.000[a]
Age	Younger	Older	.000[a]
Education	Less educated	More educated	.000[a]
Sex	Slightly more females	Slightly more males	.000[a]
Race/ethnicity	More minorities	More whites	.000[a]
Social-psychological variables			
Dogmatism	More dogmatic	Less dogmatic	.000[b]
Conservatism	More conservative	Less conservative	.000[b]
Political incapability	More feeling of incapability	Less feeling of incapability	.000[b]
Discontent with politics	More feeling of discontent	Less feeling of discontent	.000[b]
Big business	More negative	More positive	.000[b]
Big government	More positive	Less positive	.000[b]
Communication patterns			
Spouse	Very frequently	Frequently	.000[b]
Children	Frequently	Occasionally	.000[b]
Friends	Frequently	Occasionally	.000[b]
People at work	Very frequently	Frequently	.000[b]
Neighbors	Frequently	Occasionally	.000[b]
Complainers' index	More complaints	Fewer complaints	.000[b]

[a] = Chi-square test.
[b] = F-test.
Source: Compiled by the authors.

private organizations than the less conserving group does. (See Tables D.7-D.12 for detailed data on this factor.)

Factor 3—Reduced Energy-Consuming Activities

The third factor implies greater conservation efforts than did the first two. It is made up of nine items ranging from replacing high-wattage light bulbs with lower-wattage ones to hanging clothes to dry, defrosting the freezer more often, and replacing appliances with more efficient ones (see Table 4.8). On this factor 18 percent of the sample had scores indicating substantial to moderate conservation efforts and compose the more conserving group, while 82 percent had scores indicating slight to no efforts (the less energy-conserving group). These figures are exactly the opposite of those for the first factor and indicate how sharply conservation falls off as it becomes more difficult.

TABLE 4.11

Consumer Comparison on Factor 3,
Reduced Energy-Consuming Activities

	More Energy-Conserving	Less Energy-Conserving	Significance Level
Demographic variables			
Income	Lower income	Higher income	.000[a]
Age	Older	Younger	.000[a]
Education	Less educated	More educated	.000[a]
Sex	More females	Fewer females	.000[a]
Race/ethnicity	More minorities	More whites	.000[a]
Social-psychological variables			
Dogmatism	More dogmatic	Less dogmatic	.000[b]
Conservatism	More conservative	Less conservative	.000[b]
Political incapability	More feeling of incapability	Less feeling of incapability	.000[b]
Discontent with politics	More feeling of discontent	Less feeling of discontent	.000[b]
Big business	More negative	More positive	.000[b]
Big government	More positive	More negative	.000[b]
Communication patterns			
Spouse	Very frequently	Frequently	.000[b]
Children	Very frequently	Frequently	.000[b]
Friends	Very frequently	Frequently	.000[b]
People at work	Very frequently	Frequently	.000[b]
Neighbors	Frequently	Occasionally	.000[b]
Complainers' index	More complaints	Fewer complaints	.000[b]

[a] = Chi-square test.
[b] = F-test.
Source: Compiled by the authors.

The demographic profile on the more conserving group is much the same as it was for the previous factors (see Table 4.11). More than 33 percent of the under $5,000 income group is classified as more conserving. The percentages decline across income groups to a low of 8 percent among respondents with incomes of $25,000 or more. Similarly, on education the percent of respondents ranking as higher conservers drops from 38 percent to 13 percent as one goes from low to high education categories. Thirty-three percent of the black respondents and 25 percent of the Mexican-American respondents are classified in the more conserving group, compared with 16 percent of the whites. A reversal is found in terms of age, however. On factor 2 younger respondents were more heavily represented among the more conserving; but on factor 3 nearly 30 percent of those over 60 years are so classified, compared with 11 percent of the under-30 group. Although a vast majority of both males and females are classified as less conserving, a somewhat higher percentage of women are categorized as more conserving (32 percent) than men (24.8 percent).

On social-psychological variables the pattern seen on the two earlier factors is maintained. The more conserving group tends to be more dogmatic and more conservative, and to have stronger feelings of political incapability and discontent than does the less energy-conserving group. In addition, the conserving subjects are more negative toward big business and more positive toward government intervention. We continue to find, too, that the respondents who report more conservation also report more frequent discussion of energy matters. (See Tables D.13-D.18 for further information on factor 3.)

Factor 4—Installed Energy-Conserving Materials

The fourth, and last, dimension of energy conservation behavior includes five items relating to the installation of energy-conserving materials in the home: storm windows, weatherstripping, Thermopane windows, insulating drapes, and insulation (see Table 4.8 for factor loadings). On this factor 27 percent of the respondents are classified as more conserving and 73 percent as less conserving.

Somewhat surprisingly, as Table 4.12 indicates, the same profile of the more conserving group is found, although one might expect low-income respondents to report lesser efforts when dollar investments are involved. Nevertheless, nearly 50 percent of the under $5,000 income group ranks as more conserving, a figure falling to 19 percent at the highest income level (see Tables D.19-D.24 for complete data on this factor). From lowest to highest education levels, the percentage receiving higher conservation scores decreases from 44 percent to 21 percent, respectively. Respondents over age 60 continue to report the most conservation effort, with the middle-age groups having the lowest scores on this factor. More minorities and females report more energy conservation efforts than do the whites and males in the sample.

TABLE 4.12

Consumer Comparison on Factor 4, Installed Energy-Conserving Materials

	More Energy-Conserving	Less Energy-Conserving	Significance Level
Demographic variables			
Income	Lower income	Higher income	.000[b]
Age	More over 60 years	Middle-age to younger	.000[b]
Education	Less educated	More educated	.000[b]
Sex	More females	More males	.000[b]
Race/ethnicity	More minorities	White	.000[b]
Social-psychological variables			
Dogmatism	More dogmatic	Less dogmatic	.000[a]
Conservatism	More conservative	Less conservative	.000[a]
Political incapability	More feeling of incapability	Less feeling of incapability	.000[a]
Discontent with politics	More feeling of discontent	Less feeling of discontent	.000[a]
Big business	More negative	More positive	.001[a]
Big government	More positive	More negative	.000[a]
Communication patterns			
Spouse	Very frequently	Frequently	.000[b]
Children	Frequently	Frequently-occasionally	.000[b]
Friends	Frequently	Occasionally	.000[b]
People at work	More than frequently	Less than frequently	.000b
Neighbors	Frequently-occasionally	Occasionally	.000[b]
Complainers' index	No difference	No difference	n.s.[b]

[a]= Chi-square test.
[b]= F-test.
n.s. = not statistically significant.
Source: Compiled by the authors.

Each of the social-psychological variables distinguished between the more and less energy-conserving subjects. The more conserving group was more dogmatic, more conservative, and had more feelings of political incapability and discontent. They were also more negative toward big business and more positive toward big government. Finally, their communication patterns were similar to the more conserving subjects in the previous three factors. That is, the more conserving subjects were more likely to discuss the energy problem with other people than were the less conserving respondents. There was no significant difference between the groups in terms of complaints.

SUMMARY

Chapter 4 has presented a discusssion of consumers' behavior as it relates to the energy problem. It has shown that although the subjects discussed the energy problem with a number of different people, they were quite reluctant to complain to either public or private officials. The chapter has also showed that the consumers were willing to make substantial efforts to conserve energy as long as they were not forced to spend substantial sums of money or experience a negative impact on their life style. The chapter concludes with a discussion of four conservation factors. It shows that in most cases those individuals who were classified as more energy-conserving were lower-income, less educated, and more likely to be of a minority race or ethnic group than were the less energy-conserving subjects. Chapter 5 presents a discussion of the effectiveness of energy conservation incentives.

REFERENCES

Friedmann, Karl A. 1974. *Complaining: Comparative Aspects of Complaint Behavior and Attitudes Toward Complaining in Canada and Britain*. London: Sage Publications.

Gottlieb, David, and Marc Matre. 1976. *Sociological Dimensions of the Energy Crisis: A Follow-up Study*. Houston: University of Houston, Energy Institute.

Opinion Research Corporation. 1975. *Public Attitudes Toward Energy*. Highlight Report series. Princeton, New Jersey: Opinion Research Corporation.

5

ENERGY CONSERVATION
INCENTIVES FROM THE
CONSUMER'S POINT OF VIEW

Examination of consumer response to various incentives designed to encourage energy conservation is just beginning. Preliminary findings in the area are not encouraging. The response given first and foremost by consumers is that the most acceptable energy policies are those requiring the least inconvenience, the least personal cost, and the least change in life style (Doering et al. 1974; Gottlieb and Matre 1976; Grier 1976). Ted Bartell (1974) reports greatest public acceptance for behavioral regulation such as the 55 mph speed limit, a required reduction of 10 percent in electricity use, and reserved freeway lanes for buses and car pools. Economic restrictions such as added taxes on fuels are viewed with great disfavor (ORC 1975). Fuel rationing even appears to have higher priority with consumers than do higher prices in any form.

ECONOMIC INCENTIVES

Positive economic incentives—those that can help the consumer save money—may be more efficient then either behavioral regulation or price regulation, for home insulation and improvements had high consumer acceptability. Similarly, the ORC study (1975) showed that consumers view rebates from the federal government of 25 percent to 50 percent of the cost of added home insulation and storm windows as encouraging.

A major question concerns which groups of consumers would be most affected by various incentive programs—that is, who would take advantage of them. Consumers at higher income levels use the most energy, but they can also best afford the cost of additional home improvements and already have the most energy-efficient residences. Seymour Warkov (1976) found in his Houston-area sample that added home insulation was more prevalent at higher income levels. Eunice Grier (1976) pinpoints the middle-income group ($14,000-16,999) as having made the largest proportion of energy-saving home improvements.

Several studies are available that deal with the impact of price adjustments on various income sectors of society (for example, Berman and Hammer 1973; Berman, Hammer, and Tihansky 1972). The major suggestions emerging from these analyses are the not-surprising ones that high-income groups display a greater ability then low-income groups to reduce electrical consumption, that both direct and indirect energy expenditures are regressive in nature, and that the ability of low-income groups to reduce their consumption of energy is relatively low (see Newman and Day 1975, for an overview).

In summary, not very much is known about consumer reactions to conservation incentives—whether they be price, low-interest loans, or tax rebates. Nor is much known about how the consumer views a conservation investment in terms of the amount of time within which he expects to recoup his initial outlay through energy savings. It is not even known whether public policies designed to promote energy efficiency or to reduce consumption will influence the consumer at all. It may be that, as Gordon Bultena (1976) argues, consumers will continue to be essentially unchanged in their behavior for some time as they count on technological solutions to problems of energy supplies. The present chapter presents the data relevant to these questions.

RESEARCH OBJECTIVE

The specific objective of this chapter is to examine consumer reactions to various incentives that could be used to influence energy conservation behavior. The analysis includes responses to questions concerning price increases in gasoline, electricity, and natural gas, and responses involving the amount of time consumers feel is acceptable to recover investments in insulation, storm windows, and solar energy equipment. The chapter concludes with an examination of the impact of government-guaranteed, low-interest loan programs on willingness to purchase energy-saving equipment and materials.

METHOD OF ANALYSIS

For this portion of the research, the subjects have been broken down into six groups based on total annual family income: <$5,000; $5,000-9,999; $10,000-14,999; $15,000-19,999; $20,000-24,999; >$25,000. We have focused on income for several reasons. In 1976, Congress passed legislation with energy incentive provisions that can have an impact on a family's taxable income. In addition, most incentive systems are in some way based on price; it seems reasonable to focus the present research on consumer ability and willingness to respond to price mechanisms.

Analysis of variance is reported for the data on price increases in gasoline, electricity, and natural gas. This procedure is also used for analyzing the data on acceptable payback periods for insulation, storm windows, and solar energy equipment. The data on government low-interest loans is reported in percentages with the chi-square statistic only.

PRICE INCREASES IN ENERGY

The subjects were asked to state what their reactions in terms of fuel use would be to price increases for gasoline, natural gas, and electricity varying from 5 percent to more than 150 percent. The response categories were "no reduction," "slight reduction," "moderate reduction," "substantial reduction," "maximum possible reduction," and "would no longer use." The data are presented so that a score of 1 equals no reduction, a score of 2 equals a slight reduction, and so forth.

Gasoline

Figure 5.1 compares the mean responses of the total sample with the mean responses of each of the six income groups to price increases in gasoline ranging from 5 percent to 150 percent. It is apparent that even a 5 percent increase in the price of gasoline elicits some energy-conserving reactions on the part of all income groups. A 10 percent increase generates a mean response from all consumers of 1.95, just below the "slight reduction in the use of gasoline" category. A 30 percent increase in the price of gasoline elicits a mean response score of 3.01, which is virtually equivalent to "moderate reduction in the use of gasoline." A 50 percent increase elicits an average response of 3.58, which is midway between moderate and substantial reduction. It is important to note that while there is wide variation by income group, the average score for the total sample on a 100 percent price increase is 4.0, which is equivalent to "substantial reduction in the use of gasoline."

The patterns reflected in Figure 5.1 are interesting, particularly since they are basically consistent throughout the incentive analysis. In the first place, a 50 percent price increase marks a threshold level in terms of consumer response. Price increases up to 50 percent elicit markedly rising levels of reported conservation. After the 50 percent mark, however, the response levels off and little further conservation is reported. It is not possible to determine whether this pattern is due to consumer refusal to cut back on gasoline use past "substantial" reduction, or whether consumers simply find it difficult to believe that price

FIGURE 5.1

Consumers' Responses to Percentage Price Increases in Gasoline

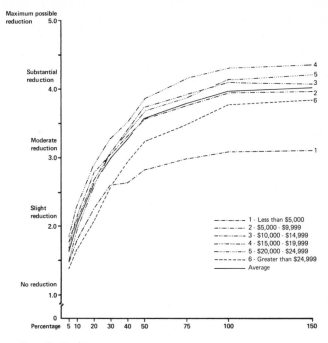

Source: Compiled by the authors.

increases up to 150 percent over present levels would occur. Average gasoline prices in the Southwest had already risen 30 percent from October 1974 to October 1975, when the survey was taken. Consumers may have viewed the hypothetical increases above 50 percent as unrealistic.

A second pattern involves the low price-responsiveness of the lowest and the highest income groups or, conversely, the high responsiveness of the middle-income consumers. This finding is consistent with earlier work (such as Grier 1976). The assumption is widespread that low-income groups cannot do much about their energy use and that high-income groups will not. The present findings tend to bear out that assumption.

Those consumers whose total family income is below $5,000 per year are by far the least price-responsive. Their conservation efforts tend to level off after the 30 percent price increase in gasoline. The gap between the responses of this group and those of the other groups is substantial. The difference between the mean response score of the lowest income group and the mean score of the sample at the 100 percent hypothetical price increase, for instance, is .91 (4.0 - 3.09). It is difficult to say why the pattern of response is not linear. That is, why is it that low-income people do not indicate "maximum" reduction in use at 30 percent price increases instead of responding with only "moderate" reduction? Any explanation here is post hoc, but it may reflect the attitude that "there is nothing more we can do." Low-income people use much less fuel than other groups of consumers, so that a moderate reduction in that use does not signifiy the same things as a moderate reduction among more affluent consumers. Dorothy Newman and Dawn Day (1975, p. 90) show that the average gasoline use (in BTU per household) of the poor is 34, compared with 85 for the lower-middle income group, 153 for the upper-middle income group, and 180 for the well-off group. Gasoline use is thus more than five times higher for the well-off than for the poor.

Consumers with family incomes above $25,000 yearly are the least price-responsive at the lowest levels of increase (5-30 percent). At the 30 percent level they become more responsive than the low-income group, but it is not until the 100 percent increase that they draw closest to the middle-income groups. At all hypothetical price levels, the consumers in the $15,000-19,999 category show the most reported behavioral change. The differences in response by income group are statistically significant (see Table 5.1) as well as substantively clear from the graphs. It is apparent that price increases are much more effective incentives for some groups of people than for others, and that the high energy users are not those most affected.

TABLE 5.1

Results of Analysis of Variance for Gasoline, Electricity, and Natural Gas Price Increases

Percent Price Increase	F-Ratio Gasoline	F-Ratio Electricity	F-Ratio Natural Gas
5	7.34	6.28	7.08
10	10.49	5.84	5.97
20	11.41	7.01	6.87
30	9.16	6.33	7.47
40	9.74	8.37	8.51
50	11.57	9.35	7.83
75	12.19	9.88	9.05
100	11.60	11.50	10.69
150	10.45	11.41	12.53

Note: All F-ratios are significant at $p < .000$ and all have five degrees of freedom.
Source: Compiled by the authors.

Electricity

The responses of consumers to price increases in electricity are quite similar to those for price increases in gasoline. Figure 5.2 indicates that the mean response for the total sample to percentage price increases in electricity is practically linear for 5 percent to 50 percent. Consumers are willing to make only the most modest conservation adjustment to a 5 percent price increase. However, they appear increasingly price-responsive up to a 50 percent price increase, which elicits a mean response for the entire sample of 3.4, substantially in excess of "modest reduction in the use of electricity." Increases above 50 percent elicit additional conservation, although the behavior curves begin to level off. For the 100 percent to 150 percent price increases, the sample mean response barely changes.

Analysis of variance significantly differentiates the income groups on the basis of responses to the various price increases (see Table 5.1). The lowest income group (<$5,000) is the most responsive to a 5 percent increase in the

FIGURE 5.2

Consumers' Responses to Percentage Price Increases in Electricity

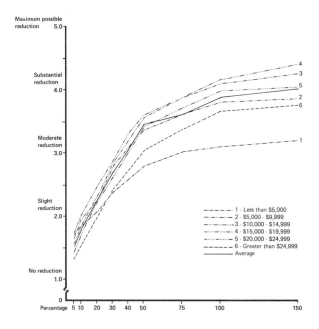

Source: Compiled by the authors.

price of electricity, with a mean score of 1.72. This pattern quickly shifts, however, with price increases in the range of 30 percent to 150 percent.

With two exceptions (the 5 percent and 75 percent price increases), the most responsive subjects to price increase were in the $15,000-19,999 income group. For both 100 percent and 150 percent price increases, the consumers in this group indicated, as a whole, that they would make more than a "substantial reduction in the use of electricity."

The patterns of response to price increases in electricity are similar to those for gasoline. The highest and lowest income groups are the least responsive to price, and the middle-income groups are the most responsive. Degree of reduction in use is virtually the same for the two fuels.

Natural Gas

Figure 5.3 presents the relationship between increases in the price of natural gas and consumer willingness to conserve that fuel. The consumer reaction pattern is similar to that for price increases in both gasoline and electricity. The mean response for a 5 percent price increase is only 1.44, less than halfway between "no reduction in the use of gasoline" (score = 1.0) and "slight reduction in the use of gasoline" (score = 2.0). Larger increases in price elicit conservation responses that are much more substantial. As an example, a price increase of 30 percent elicits a mean score for the entire sample of 2.56, while a 50 percent increase receives a score of 3.21, which is in excess of a "moderate reduction in the use of natural gas."

The curvilinear pattern that appears for gasoline and electricity also exists for natural gas. Price increments beyond 50 percent induce additional conservation efforts at a rapidly decreasing rate. A 75 percent increase receives a score of 3.47, while the 150 percent increase has a mean score of 3.93.

The analysis of variance as reported in Table 5.1 indicates that by income group the subjects are significantly differentiated at all price-increase levels. The differences within the sample are less for small price increases than for larger ones. For example, the differences between the most energy-conserving group ($15,000-19,000) at the 30 percent price level and the least energy-conserving group (<$5,000) at this level is .41, while the difference between these two income groups at the 150 percent price increase level is 1.12.

Implications

Two key findings emerge from the data presented thus far. First, conservation of gasoline, electricity, and natural gas can be achieved with price increases. While this finding is not surprising, it is important to know the price levels that are most apt to encourage conservation. In all three cases an increase of 50

FIGURE 5.3

Consumers' Responses to Percentage Price Increases in Natural Gas

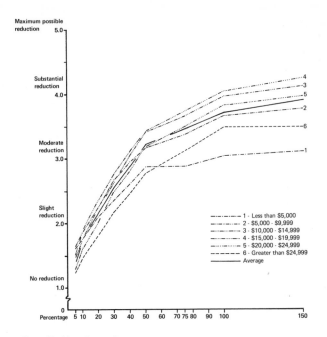

Source: Compiled by the authors.

percent appears to be the point below which substantially less conservation occurs and above which little additional conservation occurs.

Second, it is possible to identify different conservation patterns on the part of the six income groups in response to changes in the price of natural gas, electricity, and gasoline. Although each income group displayed a curvilinear reaction to the price increases, the $10,000-14,999, $15,000-19,999, and $20,000-24,999 income groups were generally the most willing to conserve energy. As an example, at price increases from 75 percent to 150 percent, these groups scored above the average for the entire sample on conservation efforts for all three fuels. In contrast, the lowest income segment, under $5,000, was normally the least willing to conserve, while the next least willing to conserve was the wealthiest segment of society. Although it is not possible to know exactly why this pattern exists, it is reasonable to speculate that the low-income consumers are already conserving energy and are not in a position to conserve more, while the upper-income people are not as concerned about the cost of energy as is the sample as a whole.

RECOVERY TIME FOR INVESTMENTS

We have seen that price increases affect various income groups differently. Another economic question of importance to the development of conservation incentives involves the amount of time acceptable to consumers for recovery of investments in more energy-efficient homes. Examination of this question can help in understanding how consumers feel about front-end costs (initial dollar outlays) relative to life-cycle savings from energy improvements.

The respondents were asked to indicate the maximum amount of time that would be acceptable to them for recovery of various levels of investments in insulation, storm windows, and solar energy equipment through savings in their bills. The response categories were "less than one year," "1-2 years," "3-4 years," "5-6 years," "7-8 years," and "more than 8 years."

Insulation

The subjects were presented with several investment capital alternatives for insulation: $100, $200, $300, $400, and $500. As Figure 5.4 indicates, the mean responses for the entire sample show a linear pattern. For an investment of $100, the sample as a whole indicates that investment should be recovered in approximately one year. Each additional $100 of investment means that investors will wait approximately eight additional months for recovery. As a result, for an investment of $500 the sample as a whole responds that investment recovery must be made in approximately four years.

It is interesting to note from Figure 5.4 that the six income groups are dichotomized in their responses. On one hand are consumers in the two lowest income segments (<$5,000 and $5,000-9,999). These consumers clearly must be able to recover even small amounts of investments in insulation very quickly. For example, the lowest income segment feels it can wait no longer than approximately six months to recover an expenditure of $100 on home insulation. For an investment of $500, this group would be willing to wait only a little longer than 1.5 years to recover the investment.

On the other hand, the four wealthiest segments of the population are rather similar in their behavior toward insulation. They are willing to accept longer periods of time to recover investments from savings in insulation. For example, the $15,000-19,999 income segment is consistently willing to wait longer than any other income group to recover the cost of insulation. They feel they need to recover an investment of $100 in just under 1.5 years, while an investment of $500 finds them willing to wait more than 4.5 years to recover their expenditures on insulation. Table 5.2 indicates that the six income groups are significantly different in terms of their responses.

FIGURE 5.4

Maximum Acceptable Time to Recover Investments in Insulation

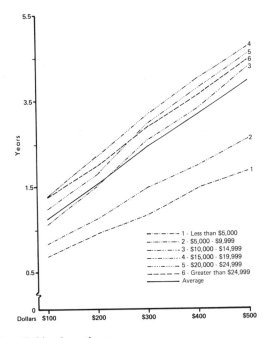

Source: Compiled by the authors.

Storm Windows

The subjects were also asked how long they would be willing to wait to recover, through savings on their energy bills, expenditures of $100, $200, $300, $400, $600, and $800 on storm windows. As Figure 5.5 indicates, the mean response follows a more curvilinear pattern than the linear pattern evidenced for insulation. As the capital requirement increases from $200 to $300 and then to $400, the subjects indicate that they would need to recover their investments in just under 1.5 years, less than 2.5 years, and less than 3.5 years, respectively. However, for an investment of $800 they indicate that they must recover the money in slightly less than 4.5 years.

There are, again, significant differences among the income groups in terms of their responses to the questions on storm window investment (see Table 5.2). Examination of Figure 5.5 shows the same dichotomy between income groups found for home insulation. The lowest income segment indicated that it must recover $100 expended on storm windows in less than one year. In contrast,

FIGURE 5.5

Maximum Acceptable Time to Recover
Investments in Storm Windows

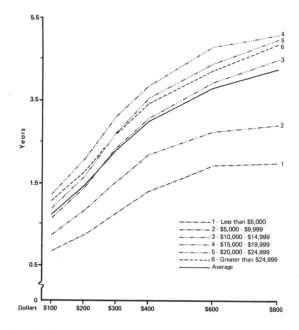

Source: Compiled by the authors.

the four highest income segments indicated they would be willing to wait more than one year.

The lowest income group must recover its outlay more quickly than any other income segment at each level of investment. In contrast, the fourth highest income segment is willing to wait the longest to recover investment in storm windows. It is also interesting to note that there are substantial differences between the extreme groups. For example, for an investment of $600 the lowest income segment feels it must recover its investment in storm windows in just over 1.5 years, while the fourth highest income segment is willing to wait more than 4.5 years to recover this expenditure.

Solar Energy

The subjects were asked what would be the maximum amount of time that they would be willing to wait to recover five levels of capital investment

TABLE 5.2

Results of Analysis of Variance for Capital Investments in Insulation and Storm Windows

Investment Alternatives	F-Ratio Insulation	F-Ratio Storm Windows
$100	13.05	9.66
200	15.85	13.05
300	19.23	17.08
400	18.95	16.42
500	20.57	—
600	—	16.83
800	—	18.91

Note: All F-ratios are significant at $p < .000$ and all have five degrees of freedom.
Source: Compiled by the authors.

in solar energy equipment through savings in energy bills. The required invest-ments were $500, $1,000, $3,000, $10,000, and $15,000. The response cate-gories again ranged from less than one year to more than eight years.

As Figure 5.6 indicates, the mean responses to the solar energy invest-ments are very curvilinear. For a $500 investment the subjects are willing to wait slightly more than 1.5 years. For a $1,000 investment the average acceptable time period is more than 2.5 years. However, as the investment increases in size the consumers are less willing to wait proportionally longer periods to recover their money. For example, on a $10,000 investment the average subject is willing to wait only slightly more than 4.5 years. For a $15,000 investment the time is only a little longer than five years.

Once again, the lowest two income segments tend to react in the same manner, while the highest four income segments also tend to respond similarly. Although this pattern of responses also was found in the examination of the storm window and insulation data, it is much more pronounced with the solar energy data.

The responses for the two lowest income groups change substantially from $500 to $1,000 to $3,000 investments. For example, the second highest income group feels it should recover its $500 investment in solar energy in slightly less than 1.5 years, while for an investment of $3,000 it is willing to wait more than 2.5 years. However, for investments from $3,000 to $15,000, the two lowest income segments do not vary greatly. The second lowest group is willing to wait only 3.5 years to recover a $15,000 investment in solar energy.

In contrast with the pattern of the two lowest income segments, the upper-income segments were more willing to wait substantially longer periods to

FIGURE 5.6

Maximum Acceptable Time to Recover Investments in Solar Energy

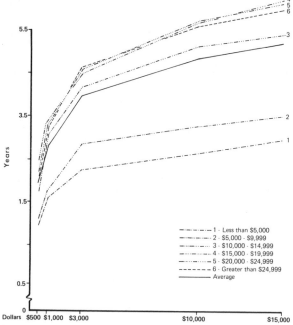

Source: Compiled by the authors.

recover their investments in solar energy equipment. As Figure 5.6 indicates, there are few differences among the response patterns for the three highest income groups ($15,000 to more than $24,000). The third highest income ($10,000 to $14,999) segment behavior is closer to the top three than the bottom two. The three highest income groups are willing to wait more than 4.5 years to recover a $3,000 investment in solar energy equipment. For an investment of $15,000, each of the three highest income segments is willing to wait more than 5.5 years. Table 5.3 indicates that there were significant differences among the income groups' responses to the solar heating and cooling investment questions.

Implications

As with responses to price increases, clear patterns emerge in responses concerning recovery time for costs of energy-saving investments. In this case, however, the two lowest income groups are distinctly split from the other groups.

TABLE 5.3

Results of Analysis of Variance for Capital Investments in Solar Energy

Investment Alternatives	F-Ratio
$ 500	7.32
1,000	13.99
3,000	15.50
10,000	22.23
15,000	23.57

Note: All F-ratios are significant at p < .000 and all have five degrees of freedom.
Source: Compiled by the authors.

Low-income consumers do not seem likely to make investments that require long payback periods. The middle-income and upper-income groups are more willing to wait longer to recover their investment. Even for large investments, however (on solar energy equipment, for example), consumers generally expect a recovery time of less than six years. This seems to indicate that long-term, life-cycle energy savings from home improvements will not be effective inducements and that more direct approaches must be made to lower initial investments.

GOVERNMENT LOANS

One way to reduce the investment cost of home energy improvements for the consumer is through government-guaranteed, low-interest loans. The sample respondents were asked to indicate whether such loans would affect their decision to purchase insulation, storm windows, or solar heating and cooling equipment. Three response categories were offered: "should convince me," "would encourage me," and "would have no effect on me."

Insulation

The data reported in Table 5.4 show few differences among income groups in their responses to low-interest loans for home insulation. The highest income group was the most apt to report that the loans would have no effect on their decision to purchase insulation. For the sample as a whole, approximately 12 percent thought the loan would convince them to add insulation, 43 percent said it would encourage them, and 45 percent reported it would have no effect. Government loans appear to be most effective for the middle-income categories ($5,000 to $19,999). This finding repeats earlier patterns—each incentive mechanism examined here seems to get the most response from these consumers.

Storm Windows

The response pattern with regard to loans for the installation of storm windows, if needed, is similar to that for insulation. Of the whole sample, 10 percent reported such loans would convince them to add storm windows, 40 percent said they would be encouraged, and 50 percent felt the loans would have no effect on their decision. In this case the effect on the higher income group appears to be even less than before (see Table 5.5). It may be that insulation is perceived as a better energy investment than storm windows, because of the widespread advertising from the former. It is important, however, that 50 percent of the sample would at least be encouraged to purchase storm windows if low-interest loans were available for that purpose.

Solar Heating and Cooling Equipment

Patterns of response for purchase of solar heating and cooling equipment show that government-guaranteed loans would have even more effect than on insulation or storm window purchase. Nearly 20 percent of the sample said they would be convinced to make the purchase and 48 percent reported they would be encouraged. It may be that larger investment outlays are anticipated for solar equipment, and that loans would be required more frequently.

Anticipation of large investments that they cannot make even with such loans may be the reason why the lowest income group has the highest "no effect" response (see Table 5.6). Once again, the middle-income groups are the most responsive to the loans.

Implications

For each of the three incentive-related situations—price increases, payback periods, and government-guaranteed, low-interest loans—it is the middle-income consumer who is reached. The low-income consumer appears to be unable to react to the incentives, while the high-income consumer, we hypothesize, will not do so. In the long run, this could lead to ever-increasing polarization by social class in American society. As Newman and Day (1975, p. 90) note, "The energy gap will begin to approximate the income gap, and the very rich will tower over all with their private yachts, executive jets, and multiple homes

TABLE 5.4

Impact of Guaranteed Low-Interest Loans on the Decision to Purchase Insulation
(percent)

	Income Group					
	<$5,000	$5,000–9,999	$10,000–14,999	$15,000–19,999	$20,000–24,999	≥$25,000
Would con-vince me	13.1	13.1	11.8	12.8	9.0	10.8
Would encourage me	41.3	45.4	46.1	46.7	41.5	34.1
Would have no effect on me	45.6	41.5	42.1	40.5	49.5	55.0
Percent	100.0	100.0	100.0	100.0	100.0	100.0
n	160	357	499	452	299	325

Note: Chi-square value = 25.25, with 10 degrees of freedom; significant at P <.005.
Source: Compiled by the authors.

TABLE 5.5

Impact of Guaranteed Low-Interest Loans on the Decision to Purchase Storm Windows
(percent)

	Income Group					
	<$5,000	$5,000–9,999	$10,000–14,999	$15,000–19,999	$20,000–24,000	≥$25,000
Would con-vince me	11.3	11.5	11.8	8.4	7.8	8.8
Would encourage me	44.6	40.9	41.5	44.7	37.6	32.1
Would have no effect on me	44.1	47.6	46.7	46.9	54.6	59.1
Percent	100.0	100.0	100.0	100.0	100.0	100.0
n	168	374	499	454	306	340

Note: Chi-square value = 25.93, with 10 degrees of freedom; significant at p <.004.
Source: Compiled by the authors.

TABLE 5.6

Impact of Guaranteed Low-Interest Loans on the Decision to Purchase Solar Heating and Cooling Equipment
(percent)

	Income Group					
	<$5,000	$5,000–9,999	$10,000–14,999	$15,000–19,999	$20,000–24,999	≥$25,000
Would con-vince me	19.5	21.6	22.4	18.9	18.6	15.6
Would encourage me	35.4	41.5	47.7	53.5	50.7	52.9
Would have no effect on me	45.1	36.9	24.9	27.5	30.7	31.5
Percent	100.0	100.0	100.0	100.0	100.0	100.0
n	164	371	495	454	306	340

Note: Chi-square value = 34.72, with 10 degrees of freedom; significant at p <.000.
Source: Compiled by the authors.

and cars." Economic-based incentives will have to be accompanied by other incentives, including social approval and direct regulation, if the high-income, high-consuming sectors of society are to be affected.

SUMMARY

Middle-income consumers appear to be most responsive to economic incentives to conserve energy. Price increases elicit the greatest conservation response from these consumers, while the lowest and highest income sectors show significantly less response. Guaranteed loan programs seem to be most encouraging to the middle-income group. It is this group, as well, that is willing to wait longer for payback of home improvement investments. It seems clear that different kinds of energy policies will be required to elicit conservation from upper-income consumers. These considerations are taken up in Chapter 6.

REFERENCES

Bartell, Ted. 1974. "The Effects of the Energy Crisis on Attitudes and Life Styles of Los Angeles Residents." Paper presented at the 69th Annual Meeting of the American Sociological Association, Montreal (August).

Berman, M.B., and M.J. Hammer. 1973. *the Impact of Electricity Price Increases on Income Groups: A Case Study of Los Angeles.* Santa Monica, Calif.: The Rand Corporation. R-1102 NSF/CSA.

Berman, M.B.; M.J. Hammer; and D.P. Tihansky. 1972. *The Impact of Electricity Price Increases on Income Groups: Western United States and California.* Santa Monica, Calif.: The Rand Corporation. R-1050 NSF/CSA.

Bultena, Gordon L. 1976. *Public Response to the Energy Crisis: A Study of Citizens' Attitudes and Adaptive Behavior.* Ames: Iowa State University.

Doering, Otto C.; Jerry Fezi; Dave Gauker; Mike Michaud; and Steve Pell. 1974. *Indiana's Views on the Energy Crisis.* Lafayette, Indiana: Purdue University, Agricultural Economics Department, Cooperative Extension Service.

Gottlieb, David, and Marc Matre. 1976. *Sociological Dimensions of the Energy Crisis: A Follow-up Study.* Houston, Texas: University of Houston, Energy Institute.

Grier, Eunice S. 1976. "Changing Patterns of Energy Consumption and Costs in U.S. Households." Paper presented at Allied Social Science Association's meeting, Atlantic City, New Jersey (September).

Newman, Dorothy K., and Dawn Day. 1975. *The American Energy Consumer. A Report to the Energy Policy Project of the Ford Foundation.* Cambridge, Massachusetts: Ballinger.

Opinion Research Corporation. 1975. *Public Attitudes Toward Energy*. Highlight Report series. Princeton, New Jersey: Opinion Research Corporation.

Warkov, Seymour. 1976. *Energy Conservation in the Houston-Galveston Area Complex: 1976*. Houston: University of Houston, Institute for Urban Studies.

CHAPTER

6

PUBLIC OPINION AND
PUBLIC POLICY

This book began with an overview of the energy situation in the United States. Domestic consumption, it was noted, surpasses domestic production, with the results that prices and dependence on foreign oil imports are rising. This conjunction of events has resulted in an atmosphere of uncertainty and ambiguity with regard to a national energy policy. Consumers find it difficult to sacrifice consumption and to conserve energy when they perceive confusion and conflict among policy makers as to the causes of energy shortages. Private industry finds energy-related investment decisions increasingly frought with uncertainty, given the stop-and-go sentiments in Congress on tax incentives, pricing policies, federal land leasing, and environmental regulation. Congress, for its part, grows ever more defensive about the state of a national energy policy and attempts to form, piecemeal, some semblance of national guidance, weaving its way through the interest groups from industry, lower-level governments, and citizens—sometimes clinging to the notion of "national interest" when all else fails.

Energy policy is, in fact, probably the most difficult issue faced by the country in many decades, primarily because the interests involved are extremely convoluted and interrelated with other issues. The observation that energy underlies all other facets of life is painfully obvious: energy means production, employment, income; it also means pollution, extensive land use, and voracious water consumption. Little is known about citizen preferences to guide policy development. It was in this context that the study described and analyzed here was undertaken. In this chapter the earlier findings on consumer attitudes and behavior will be summarized, with an eye to pointing out how they may be useful from a policy perspective.

SUMMARY AND IMPLICATIONS

Attitudes

The attitudinal analysis, presented in Chapter 3, shows that a majority of southwestern consumers, as represented by this sample, believe that an energy problem exists in the United States—though they may disagree on its causes. Consumers generally seem to realize that such major energy sources as oil and gas are being depleted; but they tend to think that others, such as coal, can be developed, given appropriate technology.

Even though a majority of respondents report that they believe that the United States has an energy problem, the younger, better-educated, and higher-income consumers are most heavily represented. The analysis shows, however, that representation of various categories of consumers depends on the type of question asked. When the attitudinal questions were factor-analyzed into six issue areas, it was found that respondents with higher education and income are more likely to believe in an energy problem in general terms—that is, in terms of resource depletion, price not being a solution, and consumer responsibility. When the issue concerns whether the problem will persist over the next 5 to 20 years and whether one's own budget has been strained, then more low-to-middle-income consumers agree. Lower-class and middle-class respondents also are more willing to attribute responsibility for the problem, and to feel that energy industries have taken advantage of the situation. Middle-class respondents are particularly unhappy about what has been done to solve the nation's energy problem. On all dimensions, it was found that younger consumers voice more concern about energy matters.

The data clearly indicate that a vast majority of the consumers in the Southwest feel that private energy corporations have taken advantage of the nation's energy problem and have acted in an unethical manner. This feeling is pervasive across all social sectors, and it seems clear that strong efforts are needed on the part of the private sector both to help alleviate the energy problem and to communicate more effectively to the American public what is being done to solve the problem. In the absence of these efforts, it is not unreasonable to expect public pressure for Congress to take a much stronger regulatory stance toward the major energy corporations.

Behavior

Several aspects of energy-relevant behavior were examined in Chapter 4. Consumers rely heavily, as might have been expected, on the mass media (newspapers, television, news magazines) for energy information. According to consumers' own reports, the energy problem is extensively discussed. Approximately 70 percent of the sample respondents report such discussions with their spouses either frequently or very frequently. Sixty percent of the consumers questioned say that they have discussions at least frequently with people at work or with friends. Discussion seems to occur least with children and neighbors.

Relatively few respondents report complaining about the energy problem. Nearly 30 percent of the consumers have complained at least once to gas companies and electric companies, but only 18 percent have complained to a congressman. No significant differences were found between "complainers" and "noncomplainers" in terms of income, education, age, or sex. Slightly more blacks and Mexican-Americans fell in the "complainers" category. The pattern of the attitudinal factors, however, indicates the presence of substantial numbers of respondents with low-to-middle income and education among the complainers. That is, the types of beliefs found to be associated with such respondents in Chapter 3 are seen here for complainers. Belief in an energy problem as measured by factor 1 is less among complainers, but greater as measured by factor 2. Complainers are also more willing to assign blame for the problem, and more apt to feel that energy industries have taken advantage of the situation and that insufficient efforts have been made to solve the problem.

A large percentage of the subjects indicate that they discuss the energy problem frequently, yet relatively few of them indicate that they have complained to either public or private officials. While the energy problem may be an interesting topic of conversation, it seems that most people do not yet feel it is worth getting upset over and, as a result, do not complain to officials about it. Many people may feel that there is nothing they can do about the energy problem, so why bother to complain. For a national move toward conservation to occur, people must feel they are involved and that their opinions and ideas are heard. As a result, it is important to establish mechanisms that the average citizen can use to communicate his attitudes about the energy problem to public and private officials.

In terms of reported conservation, a majority of consumers are making at least some efforts. The more time-consuming or expensive a particular conservation item, the less response is given. Conservation questions were factor-analyzed into four dimensions of conservation behavior. On the first and easiest factor, nearly 84 percent of the respondents claim to have made substantial or moderate efforts to conserve. This figure falls to 18 percent on the factor that included a number of time-consuming items, such as hanging clothes to dry or defrosting the freezer more often. On all four factors, more consumers with lower income and education claim to have made substantial or moderate conservation efforts.

It is important to note that the data on attitudes indicate that those individuals in higher socioeconomic groups are somewhat more likely to believe the nation is facing a legitimate energy problem. However, the individuals who are making sacrifices to conserve energy appear more frequently to be people in lower socio-economic groups. These latter individuals are frustrated because they are not apt to believe that a true energy problem exists, yet they are forced to lower their standards of living as a result of the high cost of energy.

Although consumers at all income levels are willing to make some conservation efforts, they are not as likely to do so if those efforts require them to expend substantial sums of money or lower their quality of life. As a result, the government must seriously consider developing plans for incentive systems that focus not only on the economically deprived sector of our society but also on the middle and upper income groups, and that are appropriate for the type of conservation effort required.

Incentives

As reported in Chapter 5, consumer reactions to hypothetical price increases in gasoline, electricity, and natural gas fit a curvilinear pattern. Cutbacks in use increase rather rapidly as price goes up to about 50 percent, then level off, with little additional conservation occurring at greater price increases. The lowest income group (less than $5,000 per family per year) is the least price-responsive. This group of consumers reports no more than moderate reduction in use at any price for any of the three fuels. It was suggested earlier that these respondents' consumption probably is already minimal; hence, a slight to moderate reduction would be all they could do. The highest income group ($25,000 and over) is the next least price-responsive, with middle-income groups reporting the most reduction in fuel use as price goes up.

Stronger conservation reactions to price increases may not have been given because, while energy prices have risen, and continue to rise, the increases are not disproportionately high relative to price increases for other commodities and services. If inflation continues at present levels, it may be necessary to key a tax on energy to an index of inflation. Such a policy would force energy to be relatively expensive, even though other prices, and incomes, are rising as well.

The two lowest income categories separate distinctly from the rest on the amount of time that consumers would find acceptable to recover, through savings on their energy bills, investments of varying amounts in home insulation, storm windows, and solar heating and cooling equipment. On insulation these consumers report that they must have a payback period of less than a year for a $100 investment and of less than two years for a $500 investment. Among the middle to higher income groups, acceptable payback periods for the same investments are as high as a little over a year to four years. On storm windows the lowest income group reports needing a payback period of 1.5 years even for an investment of $800, whereas the higher income groups will allow up to four years. On solar equipment the investment of $15,000 still elicits a response of less than two years' payback from the lowest income group, compared with nearly six years for higher income levels. In addition to the obvious differences among income groups it is interesting to note that all groups wish to have fairly brief payback periods, even for substantial investments.

The pattern of responses on payback period varies substantially with the product in question. For example, for investments in insulation there is almost a linear relationship between the size of the investment and the amount of time consumers are willing to wait to recover that sum. For each additional $100 expenditure, consumers seem willing to wait approximately eight months to recover it. In contrast, for storm windows there is a curvilinear pattern such that consumers are willing to wait an average of almost 1.5 years to recover the first $200 investment, but want a payback period of less than four year for an $800 investment. This pattern is even more pronounced for solar energy equipment, where a payback period between 3.5 and 5 years is wanted, regardless of whether the investment is $3,000 or $15,000. It was found, in short, that the response to investment decisions differs by income groups, and that responses vary by the type of investment considered. Obviously, policy making should include consideration of different incentive mechanisms for different income classes and types of home improvements.

With regard to the impact of government-guaranteed, low-interest loans, 12 percent of the respondents say they would be convinced to add insulation to their homes, if needed. Forty-three percent say such loans would encourage them, while 45 percent say the loans would have no effect on them. Loans for storm windows elicit responses of 10 percent "convinced," 40 percent "encouraged," and 50 percent "no effect." On solar equipment the responses are 20, 48, and 33 percent, respectively. Differences by income group are slight, but the highest income respondents appear to be least affected by the possibility of government loans for home improvements.

These data seem to indicate again that incentives will be most effective if tailored to particular types of conservation effort. Loans appear to be most effective for high-cost improvements; more than 66 percent of the sample claim that such loans would at least encourage them to buy solar equipment. Still, at least 50 percent of the respondents feel they would be encouraged to buy insulation or storm windows. On the basis of these findings, along with the earlier data on attitudes and conservation efforts, a guess can be hazarded that if a comprehensive energy conservation plan were available, complete with a carefully designed system of incentives, consumers would be influenced toward more conservation. The problem seems to be not consumer acceptance, but the lack of a strong program.

A rather clear-cut picture emerges from these data concerning consumers in the Southwest. While respondents with high income and education appear to be more aware of certain energy issues, such as the depletion of oil and gas, these are not the people with the most involvement in and concern over the energy problem. Consumers with low to middle education and income experience more effects on budget and life style. They discuss energy issues, complain about the problem, and make more conservation efforts. At the same time, these people seem to be bitter about the role energy industries are playing and are willing for the government to take a hand in sorting things out.

From a policy perspective it would seem that two of the motivating factors relied on most heavily—educating the public and price increases—cannot alone, or even in combination, solve the need for energy conservation. The more educated a person, the more likely he or she is to understand the nature of the energy problem, but he or she is not thereby the most likely to make conservation efforts. Lower-income and middle-income consumers, hit hardest by price increases, may not be very knowledgeable about energy matters, but they are conserving. At the same time, they are hostile about the situation and are likely to become increasingly so as continued price rises make energy a luxury commodity affordable only by the well-to-do. Policy makers are indeed faced with a dilemma in trying to devise conservation strategies that are both equitable and efficacious. Survey data, together with social science theory, can contribute to the understanding of human motivation that policy makers need. In the following section the basic categories of consumer motivation—economic interests, knowledge, social reinforcement, and regulation—are discussed.

DIMENSIONS OF CONSUMER MOTIVATION

Economic Interests

Behavioral science theory has long yielded a central place to the pursuit of individual utility as the prime mover in social life. Since the time when a coinage system replaced a barter or exchange system, the index of individual utility has been conceived in monetary terms. This approach, parenthetically, is now under scrutiny, and the redefinition of "quality of life" and individual utility in other than dollar terms is receiving both popular and scholarly attention. It seems clear from this survey and the others reviewed that energy as a market commodity is of primary importance to consumers. That is, energy use is in large part price-determined.

In many of the surveys reviewed, consumers report that cost is the primary motivating factor behind conservation (for instance, Gottlieb and Matre 1976; Perlman and Warren 1975; ORC 1975). Economic behavior has traditionally been viewed as logical behavior. That is, it is assumed that consumers evaluate the relative prices of commodities and, on that basis, make alternative purchasing decisions. This "rational man" model of economic behavior has never been an accurate one because of impinging factors such as are described in the categories below. As a simple theoretical assumption, however, let us state that as price rises, demand falls. This assumption translates into energy conservation through higher energy costs—either directly, through free-market prices, or indirectly, through some type of energy-use taxes.

Examination of consumer response to various economic incentives designed to encourage energy conservation is just beginning. Preliminary findings in the area are not encouraging. The response given first and foremost by consumers is that the most acceptable energy policies are those that require the least inconvenience, the least personal cost, and the least change in life style. We have seen in this analysis that reported conservation efforts are greatest when those efforts are easiest to accomplish. Ted Bartell (1974) reports greatest public acceptance for behavioral regulation, such as the 55 mph speed limit, a required reduction of 10 percent in electricity use, and reserved freeway lanes for buses and car pools. Economic restrictions such as added taxes on fuels, however, are viewed with great disfavor (ORC 1975). Fuel rationing appears to have higher priority with consumers than do higher prices in any form.

Positive economic incentives—those that can help the consumer save money—may be more efficient than either behavioral regulation or price regulation (Gallup 1976). James Zuiches (1975) found that tax deductions for home insulation and improvements had high consumer acceptability. Similarly, the ORC study (1975) showed that consumers view rebates from the federal government of 25 to 50 percent of the cost of added home insulation and storm windows as encouraging. The present study found that low-interest loans would encourage home improvements.

A major question concerns which groups of consumers would be most affected by various incentive programs—that is, who would take advantage of them. Consumers at higher income levels use the most energy, but they can also best afford the cost of additional home improvements and already have the most energy-efficient residences, Seymour Warkov (1976) found in his Houston-area sample that added home insulation was more prevalent at higher income levels. Eunice Grier (1976) pinpoints the middle-income group ($14,000-16,999) as having made the largest proportion of energy-saving home improvements. We have found that low-to-middle-income consumers more frequently report conservation efforts, and are generally more influenced by conservation incentives.

Knowledge

Higher education implies both higher levels of belief in an energy problem and greater awareness of what can be done to conserve energy. Since high education and high income are correlated, however, we find that despite awareness of energy problems among better-educated individuals, their economic interests may not be such as to motivate conservation behavior. Nevertheless, consumers can act only on the basis of the knowledge they have, and survey data have shown several ways of improving levels of energy knowledge among the public. This category encompasses two levels of knowledge: individual-specific feedback on consumption and broader energy information programs.

Survey data have shown that conservation is more likely to occur if and when the energy problem is "real" for the individual consumer, and that consumers make heavy use of mass media information sources. In order to bring these factors together, the message conveyed through mass media must be that the consumer faces personal effects of the energy problem and that he can do something about it. Julie Honnold and L.D. Nelson (1976) have shown that high consumers avoid contact with information on resource scarcity. A corollary of this would be that mass media programs will be less than effective for these people, since they can more easily "select out" the programs they wish not to see. One way to enhance information transfer is to use small, local groups as dissemination points. Mancur Olsen (1965) argues that federations of small, existing groups must be used to relay selective incentives if mobilization toward a social goal is required of a large population. This is precisely the case with energy conservation for the American people.

Social Reinforcement

Humans are characterized by a sociality sentiment that provides a basis for social life and the cohesion of societies. Particularly important aspects of this sentiment are the sense for authority or leadership, the need for group approval, and the need for order (the desire to see logical relationships). Leadership has figured as an important variable in many a social theory. S.N. Eisenstadt (1971) argues that leadership, in the form of charismatic groups or personalities, is critical for the definition and implementation of social goals. In explanation of this need for authority, Joseph Lopreato (1975, p. 17) notes, "The appearance of a big politician (even at his lowest point in popularity) at the local airport or auditorium will bring many a 'skeptic' purring."

Social change always requires the backing of leaders, and such is the case with the proposed implementation of "conservation" as a social ethic. Local-level authorities and leaders can play a role in the process, another reason for the dissemination of information through small groups; but the most important level of leadership is the national one. Until Congress, for example, is ready to clearly promote conservation, individual acceptance of that goal will be minimal.

Another aspect of human sociality has to do with the need for group approval, or the "deference the individual feels for the group of which he is a part, or for other groups, and his desire to have their approval or admiration" (Pareto 1963, sec. 1153). Yet another reason for the use of small, local groups for the encouragement of conservation is found here. David Sears et al. (1976) have, in fact, found that local sanction of conservation has more effect on behavior than do broad political attitudes the individual might hold. Donald Warren and David Clifford (1975) conclude that the roles of attitudes and economics in predicating conservation behavior are valid only in the context of neighborhood and community variables. In general, they state, conservation strategies will be more successful the more they derive from the local setting.

The desire to gain social approval is in part responsible for the phenomenon known as "imitation." Social changes generally travel down the social class structure rather than up, for the simple reason that individuals are prone to imitate their superiors, not their equals or inferiors. Groups with upper incomes and more education, however, are not practicing energy conservation; and the spread of a conservation "ethic" is thus considerably slowed.

Once innovations begin to catch on—through the influence of leaders and social approval—another human need propels these innovations into institutionalization. We are referring to the need for order. In general, humans feel a need to perceive their individual goals and activities as part of a larger social order. People seek, in fact, organizational and symbolic order (Eisenstadt 1971). Once order is established or perceived, people desire conformity to that order. A tremendous push evolves toward keeping all group members in conformity with the established social goals. Full reign can be given to this need for order only when the individual feels some attachment and responsibility for seeing a given goal pursued.

Finally, but not least, it should be remembered that the most powerful conduit of desired behavior is the socialization of the individual as he matures. Attitudes and knowledge of children and youth have been included in only a few studies. Gallup (1976) reports that, to pre-teens, the energy crisis is remote and confusing, and that both teens and pre-teens feel that the major responsibility for dealing with the situation belongs to adults. ORC (1975) data, however, indicate that, according to parents, schools emphasize energy conservation, at least to some degree, and that children are making attempts to conserve energy at home as a result of things learned at school. At a higher level, many colleges and universities are adding energy-related courses to their curricula. The effect of these educational programs is yet to be seen, but the possible role of formal schooling in energy education should not be underestimated, especially in light of evidence (see, for instance, Curtin 1975; Talarzyk and Omura 1975; Zuiches 1975) that education has a primary effect on energy beliefs and conservation behavior. In the present analysis, younger respondents (generally under 30 years of age) are found to be greatly concerned about the energy problem.

Regulation

Whether prescriptive or proscriptive, regulation of behavior by some supra-individual authority will motivate behavior. This category is particularly important for energy conservation because energy use is determined in large part by the efficiencies of appliances, buildings, and materials. Energy-efficient goods must be available to the consumer, and regulation of production standards helps guarantee that availability. Much can be done in terms of policy to give consumers a more energy-efficient environment. The industrial ploy sometimes referred to as "planned obsolescence," for instance, should be discouraged in favor of longer-lasting, more efficient appliances and cars that do not have to be replaced every other year.

Regulation of appliance, car, and building standards is particularly important, given the fact that regulation of individual behavior does not yield stable results. The desired individual behavior is elicited under regulation only as "the least of evils" or to avoid trouble with those in power, not because the individual identifies it as a necessary part of his social order (Eisenstadt 1971). One of the more pressing research needs is the identification of consumer acceptance of regulation. Zuiches (1975) reports that the greater the energy awareness of an individual, the more likely he or she is to find a variety of policies, including regulatory ones, acceptable. Awareness leads to belief in the energy crisis, and thus to acceptance of regulation as a legitimate need. We have found in this study, however, that consumers at lower income and education levels are more favorably disposed toward government regulation.

The source of regulation must be perceived as legitimate. ORC (1975) data reveal that a majority of Americans did not follow federal government advice on thermostat settings during the winter of 1974-75, nor did they intend to do so the following winter. The reason most often given was that the federal government was not a good source of advice on energy-saving procedures. Obviously, before turning from voluntary to regulatory policies, the federal government would need to improve its image on the energy situation.

Incentives designed to motivate consumers to conserve energy, regardless of what they are, are useless without the development of certain societal prerequisites. These involve organizational structures, skills, and marketing and planning procedures necessary to implement conservation.

> Even assuming that certain goals have, by some calculus or consensus, been accepted within large parts of a population, such acceptance does not mean that the organizational needs and prerequisites necessitated by the implementation of these goals are automatically taken care of, or that the necessary organizations and arrangements will automatically develop and become institutionalized (Eisenstadt 1971, p. 53).

Examples relevant in the present context would be marketing and service groups for such energy innovations as solar equipment, finance strategies allowing home improvement that would decrease energy use, public transportation instead of the private automobile, or reliable sources of advice for correcting energy-use habits. Even if the individual is highly motivated to conserve, he can do little without appropriate institutional arrangements to support him.

The results of this analysis indicate widespread acceptance in the Southwest of the fact that the United States has, and will continue to have, an energy problem. Widespread, if not always effective, efforts at conservation are also indicated. Under these circumstances, failure to move forward with national energy plans and programs will mean failure to address a problem that has become a major social issue.

REFERENCES

Bartell, Ted. 1974. "The Effects of the Energy Crisis on Attitudes and Life Styles of Los Angeles Residents" Paper presented at 69th Annual Meeting of the American Sociological Association, Montreal (September).

Curtin, Richard T. 1975. "Consumer Adaptation to Energy Shortages." Ann Arbor: University of Michigan, Survey Research Center. Unpublished.

Doering, Otto C.; Jerry Fezi; Dave Gauker; Mike Michaud; and Steve Pell. 1974. *Indiana's Views on the Energy Crisis*. Lafayette, Indiana: Purdue University, Agricultural Economics Department, Cooperative Extension Service.

Eisenstadt, S.N. 1971. "Societal Goals, Systemic Needs, Social Interaction, and Individual Behavior: Some Tentative Explorations." In *Institutions and Social Excange*, ed. H. Turk and R.L. Simpson, pp. 36-55. New York: Bobbs-Merrill.

Gallup Organization, Inc. 1976. *Group Discussions Regarding Consumer Energy Conservation*. Washington, D.C.: Federal Energy Administration.

Gottlieb, David, and Marc Matre. 1976. *Sociological Dimensions of the Energy Crisis: A Follow-up Study*. Houston: University of Houston, Energy Institute.

Grier, Eunice S. 1976. "Changing Patterns of Energy Consumption and Costs in U.S. Households." Paper presented at Allied Social Science Associations meeting, Atlantic City, New Jersey (September).

Honnold, Julie A., and L.D. Nelson. 1976. "Voluntary Rationing of Scarce Resources: Some Implications of an Experimental Study." Paper presented at Annual Meeting of the American Sociological Association, New York (August).

Lopreato, Joseph 1975. *The Sociology of Vilfredo Pareto*. Morristown, New Jersey: General Learning Press.

Olsen, Mancur. 1965. *The Logic of Collective Action*. Cambridge, Massachusetts: Harvard University Press.

Opinion Research Corporation. 1975. *General Public Attitudes and Behavior Regarding Energy Saving*. Highlight Report, X. Princeton, New Jersey: Opinion Research Corporation.

Pareto, Vilfredo. 1963. *Treatise on General Sociology*. New York: Dover Press.

Perlman, Robert, and Roland L. Warren. 1975b. *Energy Saving by Households of Different Incomes in Three Metropolitan Areas*. Waltham, Massachusetts: Brandeis University, Florence Heller Graduate School for Advanced Studies in Social Welfare.

Sears, David O.; Tom R. Tyler; Jack Citrin; and Donald R. Kinder. 1976. "Political System Support and Public Response to the 1974 Energy Crisis." Paper presented at Conference on Political Alienation and Political Support, Palo Alto, California (May).

Talarzyk, W. Wayne, and Glenn S. Omura. 1975. "Consumer Attitudes Toward and Perceptions of the Energy Crisis." In *American Marketing Association 1974 Combined Proceedings*, ed. Ronald C. Curhan, pp. 316-22. Ann Arbor, Michigan: Xerox University Microfilms.

Warren, Donald I., and David L. Clifford. 1975. *Local Neighborhood Social Structure and Response to the Energy Crisis of 1973-74.* Ann Arbor: University of Michigan, Institute of Labor and Industrial Relations.

Warkov, Seymour. 1976. *Energy Conservation in the Houston-Galveston Area Complex: 1976. Preliminary Results from the Houston Metropolitan Area Project.* Houston: University of Houston, Institute for Urban Studies.

Zuiches, James J. 1975. *Energy and the Family.* East Lansing: Michigan State University, Department of Agricultural Economics.

College of Business Administration Foundation
The University of Texas at Austin
Austin, Texas 78712
Center for Energy Studies

Dear Austin Resident,

The Center for Energy Studies at The University of Texas at Austin has been involved in a number of research projects related to our nation's energy problems. At this time we would like to ask you to complete a questionnaire which involves determining your attitudes and behavior toward a variety of energy-related issues.

Although the questionnaire looks rather long, it should only take approximately 20 minutes of your time to complete. We have enclosed a postage-paid envelope for your use to return the questionnaire. There are no right or wrong answers. It's your opinion that counts. Your answers will be confidential. No one will be able to associate your name with the answers you give.

If you are married, the questionnaire can be filled out by either you or your spouse. If you are single, please fill out the questionnaire yourself.

We can't force you to help, and we can't afford to give you a reward. But we believe that in this country every person's opinion should be considered. If you will complete this questionnaire, then you will have your say on an important matter which affects us all. The results of the research will be sent to Washington and hopefully they will play a role in determining energy policy.

Dr. William H. Cunningham
Coordinator of Commercialization

Dr. Sally Lopreato
Director of Social Systems Analysis

I. We would like your opinion on the following statements. Please indicate whether you "strongly agree," "agree," "have no opinion," "disagree," or "strongly disagree" with each statement. There are no right or wrong answers; please mark the box that most closely matches what you think. As an example, if you "disagree" with the statement listed below, you would mark the answer "disagree."

	Strongly Agree	Agree	Have No Opinion	Disagree	Strongly Disagree

EXAMPLE: Most Americans have attempted to car-pool to save energy.
— — — — —

PLEASE PROCEED WITH EACH STATEMENT

1. The United States currently has an energy problem. — — — — —
2. The United States is running out of natural gas. — — — — —
3. The United States is running out of oil. — — — — —
4. The United States is running out of coal. — — — — —
5. Our energy problem exists only because we are being charged high prices for energy. — — — — —
6. President Ford is responsible for the energy problem. — — — — :—
7. The United States Congress is responsible for the energy problem. — — — — —
8. The Arab oil-exporting countries are responsible for the energy problem. — — — — —
9. President Nixon was responsible for the energy problem. — — — — —
10. The natural gas companies are responsible for the energy problem. — — — — —
11. The petroleum companies are responsible for the energy problem. — — — — —
12. The electrical-producing companies are responsible for the energy problem. — — — — —
13. The state government is responsible for the energy problem. — — — — —
14. The Israelis are responsible for the energy problem. — — — — —
15. United States consumers are responsible for the energy problem. — — — — —
16. Congress has done all it can to solve the energy problem. — — — — —
17. President Ford has done all he can to solve the energy problem. — — — — —
18. The natural gas companies did not cause the energy problem, but have taken advantage of it to raise prices. — — — — —

	Strongly Agree	Agree	Have No Opinion	Disagree	Strongly Disagree

19. The natural gas companies did not cause the energy problem, but have taken advantage of it to raise prices. __ __ __ __ __

20. The natural gas companies have raised their prices only to cover their added cost and have not profited by the energy shortage. __ __ __ __ __

21. The natural gas companies have been unfairly blamed for the energy problem. __ __ __ __ __

22. The petroleum companies have done all they could to solve the energy problem. __ __ __ __ __

23. The petroleum companies did not cause the energy problem, but have taken advantage of it to raise prices. __ __ __ __ __

24. The petroleum companies have raised their prices only to cover their added cost and have not profited by the energy shortage. __ __ __ __ __

25. The petroleum companies have been unfairly blamed for the energy problem. __ __ __ __ __

26. The electric companies have done all they could to solve the energy problem. __ __ __ __ __

27. The electric companies did not cause the energy problem, but have taken advantage of it to raise prices. __ __ __ __ __

28. The electric companies have raised their prices only to cover their added cost and have not profited by the energy shortage. __ __ __ __ __

29. The electric companies have been unfairly blamed for the energy problem. __ __ __ __ __

30. The failure of technology to develop new sources of energy is responsible for the energy problem. __ __ __ __ __

31. I feel enough information has been made available to me concerning the energy problem. __ __ __ __ __

32. The energy problem has put a substantial strain on my budget. __ __ __ __ __

33. I feel the energy problem will cause major difficulties for the United States during the next five years. __ __ __ __ __

34. I feel the energy problem will cause major difficulties for the United States during the next 20 years. __ __ __ __ __

35. There will be no energy shortage in the United States as long as we are willing to pay a high price for energy. __ __ __ __ __

II. Now we would like to ask you some questions about *sources of energy information* and *whom you talk with about energy issues.*

1. Where do you get most of your information about energy matters? Please indicate your first, second, and third most important sources by placing 1, 2, and 3 in front of those sources.

_newspapers   _radio   _friends and family
_television   _news magazines   _government literature
_high school courses   _college courses   _other (specify) _____

2. Please indicate how frequently you have discussed the energy problem with the following people. The responses range from "very frequently" to "not at all."

	Very frequently	Frequently	Occasionally	Not at all
Spouse				
Children				
Friends				
People at work				
Neighbors				

3. How often during the last two years have you complained about energy problems to your Congressman, utility companies, newspapers, state or federal officials? Please respond separately to each category.

	I have not complained during the last 2 yrs.	I complained once	I complained 2 or 3 times	I complained 4 or 5 times	I complained more than 5 times
Congressman					
Gas company					
Electric company					
Petroleum company					
Newspapers					
State officials					
Federal officials					
Other (please specify)					

III. The following questiona are about your family's energy bills.

1. Please check the space which indicates the total amount of what you believe your natural gas, electricity, and gasoline bills were during *January of 1975.*

	No Expenditure	$.01-5.00	$5.01-10.00	$10.01-20.00	$20.01-30.00	$30.01-50.00	$50.01-75.00	$75.01-100.00	More Than $100.00
Natural gas									
Electricity									
Gasoline									

2. How do you believe your *January 1975* bills compared with your *January 1974* bills for natural gas, electricity, and gasoline?

	Lower	Same	$.01-10.00 Higher	$10.01-25.00 Higher	$25.01-50.00 Higher	$50.01-75.00 Higher	More Than $75.00 Higher
Natural gas							
Electricity							
Gasoline							

3. Please check the space which indicates the total amount of what you believe you natural gas, electricity, and gasoline bills were last month.

	No Expenditure	$.01-5.00	$5.01-10.00	$10.01-20.00	$20.01-30.00	$30.01-50.00	$50.01-75.00	$75.01-100.00	More Than $100.00
Natural gas									
Electricity									
Gasoline									

4. How do you believe your last month's bill compared with your bill at the same time last year for natural gas, electricity, and gasoline?

	Lower	Same	$.01-10.00 Higher	$10.01-25.00 Higher	$25.01-50.00 Higher	$50.01-75.00 Higher	More Than $75.00 Higher
Natural gas							
Electricity							
Gasoline							

5. If your electric bills have gone up, how has that affected your family?
    ___ It really had no effect on us.
    ___ We had to make a few adjustments, but our style of life was not affected.
    ___ Our life was less comfortable and convenient, but it was not serious.
    ___ We had to make serious changes in our daily habits.
6. If your natural gas bill has gone up, how has this affected your family?
    ___ It really had no effect on us.
    ___ We had to make a few adjustments, but our style of life was not affected.
    ___ Our life was less comfortable and convenient, but it was not serious.
    ___ We had to make serious changes in our daily habits.
7. If your gasoline bills have gone up, how has this affected your family?
    ___ It really had no effect on us.
    ___ We had to make a few adjustments, but our style of life was not affected.
    ___ Our life was less comfortable and convenient, but it was not serious.
    ___ We had to make serious changes in our daily habits.

IV. 1. Which of the following measures have you taken to reduce your energy consumption? Please indicate with a check mark in the appropriate column if you have made "substantial efforts," "moderate efforts," "slight efforts," or "no effort" in each of the following areas. If a question is not applicable to you—for example, you live in an apartment and therefore you would not consider adding insulation to your roof—please check the column "not applicable."

EXAMPLE: Used attic fan less.

PLEASE PROCEED WITH EACH STATEMENT

	Substantial Efforts	Moderate Efforts	Slight Efforts	No Effort	Not Applicable
1. Tried to always turn out lights when not needed.	—	—	—	—	—
2. Turned thermostats down in winter.	—	—	—	—	—
3. Turned thermostats up in summer.	—	—	—	—	—
4. Replaced light bulbs with bulbs of lower wattage.	—	—	—	—	—
5. Used dishwasher less.	—	—	—	—	—
6. Watched TV less.	—	—	—	—	—
7. Replaced appliances with more energy-efficient ones.	—	—	—	—	—
8. Installed storm windows.	—	—	—	—	—
9. Installed weatherstripping on doors and windows.	—	—	—	—	—
10. Turned down water-heater thermostat setting.	—	—	—	—	—
11. Used less hot water.	—	—	—	—	—
12. Turned dishwasher off before dry cycle.	—	—	—	—	—
13. Used fans and opened windows instead of using air conditioning.	—	—	—	—	—
14. Defrosted freezer more often.	—	—	—	—	—

	Substantial Efforts	Moderate Efforts	Slight Efforts	No Effort
15. Hung clothes to dry rather than using the clothes dryer.				
16. Installed Thermopane windows.	—	—	—	—
17. Opened drapes during the day, closed at night during the winter.	—	—	—	—
18. Purchased insulating drapes.	—	—	—	—
19. Added insulation to attic.	—	—	—	—
20. Closed fireplace damper when not in use.	—	—	—	—
21. Turned off pilot to furnace in the summer.	—	—	—	—
22. Turned off decorative yard lights.	—	—	—	—
23. Washed only full loads in clothes washer.	—	—	—	—
24. Turned heat off during the day while away in the winter.	—	—	—	—
25. Turned air conditioning off during the day while away in the summer.	—	—	—	—
26. Closed off unused rooms.	—	—	—	—

2. Please indicate how your *present* efforts to conserve natural gas, gasoline, and electricity compare with your *last year's* efforts to conserve energy.

	Conserved Substantially Less This Year	Conserved Less This Year	Conserved About the Same This Year	Conserved More This Year	Conserved Substantially More This Year
Natural Gas					
Gasoline					
Electricity					

V. 1. Please indicate by checking the appropriate space the source of energy that your family uses to power any of the following appliances in your home.

	Central Heat	Space Heater	Floor Furnace	Wall Furnace	Range	Refrigerator	Oven	Clothes Dryer	Water Heater	Central Air Conditioner	Outdoor Light	Freezer	Outdoor Grill
Natural Gas													
Electricity													
Do not own													

2. Do you own or rent the place where you live? ___ own   ___ rent
3. Is the place where you live a single residence (house) or a multiple-living establishment (apartment or condominium)?
       ___ single residence   ___ multiple-living
4. If your family rents or leases its living accommodations, is your *natural gas bill* paid by you or is it paid by the landlord?
   ___ paid by me or my family   ___ paid by my landlord   ___ we do not have natural gas
5. If you or your family rents or leases, its living accomodations, is your *electricity bill* paid by you or the landlord?
   ___ paid by me or my family   ___ paid by my landlord   ___ we do not have electricity
6. After you have paid all the normal monthly expenses, how much money does your family have left to save or spend on special items that you do not *normally* purchase?
   ___ less than $20   ___ $51-100   ___ $201-400   ___ more than
   ___ $20-50   ___ $101-200   ___ $401-700 $700
7. How many square feet is your apartment or house?
   ___ less than 500 square feet   ___ 1,251-1,500   ___ 2,001-2,500
   ___ 500-1,000   ___ 1,501-1,750   ___ 2,510-3,000
   ___ 1,001-1,250   ___ 1,751-2,000   ___ 3,001-3,500
                                                       ___ more than 3,500
8. How old is your house? _____
9. How much insulation do you have in your attic?
   ___ none   ___ 3-4 inches   ___ more than 6 inches
   ___ 1-2 inches   ___ 5-6 inches   ___ I do not know
10. Do you have blown insulation or blanket insulation in your attic?
   ___ blown insulation   ___ no insulation
   ___ blanket insulation   ___ I do not know
11. Please indicate what you believe the value of your home is.
   ___ less than $10,000   ___ $40,001-50,000
   ___ $10,001-20,000   ___ $50,001-60,000
   ___ $20,001-30,000   ___ $60,001-70,000
   ___ $30,001-40,000   ___ more than $70,000
12. At what temperature do you set your thermostat during the winter months?
   ___
13. At what temperature do you set your thermostat during the summer months? ___
14. Please specify the make, model, and year of your family's automobile(s).
       Make   Model   Year
   1. _____   _____   _____
   2. _____   _____   _____
   3. _____   _____   _____
   4. _____   _____   _____

VI. 1. Please indicate below what your reactions would be to the following percentage increases in the price of *gasoline*. That is, if you would respond to a 2% increase in the price of gasoline by not reducing your use of it, then check the space as indicated below. Please mark your reactions to the rest of the percentage increases in price of gasoline.

	No Reduction in the Use of Electricity	Slight Reduction in the Use of Electricity	Moderate Reduction in the Use of Electricity	Substantial Reduction in the Use of Electricity	Maximum Possible Reduction in the use of Electricity	Would No Longer Use Electricity
5%						
10%						
20%						
30%						
40%						
50%						
75%						
100%						
150%						
More than 150%						

2. Please indicate below what your reactions would be to *each* of the following percentage increases in the price of *natural gas*.

	No Reduction in the Use of Natural Gas	Slight Reduction in the Use of Natural Gas	Moderate Reduction in the Use of Natural Gas	Substantial Reduction in the Use of Natural Gas	Maximum Possible Reduction in the use of Natural Gas	Would No Longer Use Natural Gas
5%						
10%						
20%						
30%						
40%						
50%						
75%						
100%						
150%						
More than 150%						

3. Please indicate below what your reaction would be to *each* of the following percentage increases in the price of *electricity*.

	No Reduction in the Use of Gasoline	Slight Reduction in the Use of Gasoline	Moderate Reduction in the Use of Gasoline	Substantial Reduction in the Use of Gasoline	Maximum Possible Reduction in the use of Gasoline	Would No Longer Use Gasoline
5%						
10%						
20%						
30%						
40%						
50%						
75%						
100%						
150%						
More than 150%						

VII. 1. Assume that adding *insulation* to the attic of your home would cost between $100 and $500. Please indicate what would be the maximum amount of time that you would accept to recover, through savings in your electric or gas bills, your expenditure of $100, $200, $300, $400, or $500.

	Less Than 1 Year to Recover Expenses	1-2 Years to Recover Expenses	3-4 Years to Recover Expenses	5-6 Years to Recover Expenses	7-8 Years to Recover Expenses	More Than 8 Years to Recover Expenses
$100 investment						
$200 investment						
$300 investment						
$400 investment						
$500 investment						

2. Assume that adding *storm windows* in your home would cost between $100 and $800. Please indicate what would be the maximum amount of time that you would accept to recover, through savings in your electric or gas bills, your expenditure of $100, $200, $300, $400, $600, or $800.

	Less Than 1 Year to Recover Expenses	1-2 Years to Recover Expenses	3-4 Years to Recover Expenses	5-6 Years to Recover Expenses	7-8 Years to Recover Expenses	More Than 8 Years to Recover Expenses
$100 investment						
$200 investment						
$300 investment						
$400 investment						
$500 investment						

3. Assume that adding *solar energy equipment* to your home would cost between $500 and $15,000. Please indicate what would be the maximum amount of time that you would accept to recover your expenditure of $500 to $15,000 through savings in your electric or gas bills.

	Less Than 1 Year to Recover Expenses	1-2 Years to Recover Expenses	3-4 Years to Recover Expenses	5-6 Years to Recover Expenses	7-8 Years to Recover Expenses	More Than 8 Years to Recover Expenses
$500 investment						
$1,000 investment						
$3,000 investment						
$10,000 investment						
$15,000 investment						

4. a. If guaranteed government loans with low interest rates were available to you for financing, how would those loans affect your decision to add *insulation*, if needed, to your home?
   __ would convince me   __ would encourage me
                    __ would have no effect on me
   b. If guaranteed government loans with low interest rates were available to you for financing, how would these loans affect your decision to add *storm windows*, if needed, to your home?
   __ would convince me   __ would encourage me
                    __ would have no effect on me
   c. Finally, how would graranteed low-interest loans affect your decision to add *solar heating and cooling equipment* to your home?
   __ would convince me   __ would encourage me
                    __ would have no effect on me

5. If you were to have more insulation placed in your attic, would you do it yourself or hire a professional?
   __ do it myself __ hire a professional __ I do not know what I would do

6. Do you own a swimming pool?
   __ yes (if so please go to question 7)
   __ no (if not, please go to section VIII)

7. If you own a pool, is it: __ not heated
                                        __ heated by natural gas
                                        __ heated by electricity

8. If you own a pool, would you want to add a solar heating system to your pool if it cost $1,300 and was not unattractive?
   __ yes, I would seriously consider it   __ yes, I would probably consider it
            __ no, a solar system is not of interest to me

VIII. Your opinion on EACH of the following statements is important, regardless of whether you have thought about them or not. Please indicate by checking the appropriate line whether you "strongly agree," "moderately agree," etc. with EACH of the statements. There are no right or wrong answers; only your opinion is important.

	Strongly Agree	Moderately Agree	Have No Opinion	Moderately Disagree	Strongly Disagree
1. I prefer the practical man anytime to the man of ideas.	—	—	—	—	—
2. If you start trying to change things very much, you usually make them worse.	—	—	—	—	—
3. If something grows up after a long time, there will always be much widsom to it.	—	—	—	—	—
4. It's better to stick by what you have than to be trying new things you don't really know about.	—	—	—	—	—
5. We must respect the work of our forefathers and not think that we know better than they did.	—	—	—	—	—
6. A man doesn't really have much wisdom until he is well along in years.	—	—	—	—	—
7. No matter how we like to talk about it, political authority really comes not from us, but from some higher power.	—	—	—	—	—
8. I'd want to know that something would really work before I'd be willing to take a chance on it.	—	—	—	—	—
9. All groups can live in harmony in this country without changing the system in any way.	—	—	—	—	—
10. I believe public officials don't care much what people like me think.	—	—	—	—	—
11. There is no way other than voting that people like me can influence actions of the government.	—	—	—	—	—
12. Sometimes politics and government seem so complicated that I can't really understand what's going on.	—	—	—	—	—
13. People like me don't have any say about what the government does.	—	—	—	—	—
14. These days the government is trying to do too many things, including some activities that I don't think it has the right to do.	—	—	—	—	—
15. For the most part, the government serves the interests of a few organized groups, such as business or labor, and isn't very concerned about the needs of people like myself.	—	—	—	—	—
16. It seems to me that the government often fails to take necessary actions on important matters, even when most people favor such actions.	—	—	—	—	—

	Strongly Agree	Moderately Agree	Have No Opinion	Moderately Disagree	Strongly Disagree
17. As the government is now organized and operated, I think it is hopelessly incapable of dealing with all the crucial problems facing the country today.	—	—	—	—	—
18. In this complicated world of ours, the only way we can know what's going on is to rely on leaders or experts who can be trusted.	—	—	—	—	—
19. My blood boils whenever a person stubbornly refuses to admit he's wrong.	—	—	—	—	—
20. There are two kinds of people in this world: those for the truth and those who are against the truth.	—	—	—	—	—
21. Most people just don't know what's good for them.	—	—	—	—	—
22. Of all the different philosophies which exist in this world, there is probably only one which is correct.	—	—	—	—	—
23. The highest form of government is a democracy, and the highest form of democracy is a government run by those who are most intelligent.	—	—	—	—	—
24. The main thing in life is for a person to want to do something important.	—	—	—	—	—
25. I'd like it if I could find someone who would tell me how to solve my personal problems.	—	—	—	—	—
26. Most of the ideas which get printed nowadays aren't worth the paper they are printed on.	—	—	—	—	—
27. Man on his own is a helpless and miserable creature.	—	—	—	—	—
28. In many of our largest industries, one or two companies have too much control of the industry.	—	—	—	—	—
29. There's too much power concentrated in the hands of a few large companies for the good of the nation.	—	—	—	—	—
30. As they grow bigger, companies usually get cold and impersonal in their relations with people.	—	—	—	—	—
31. For the good of the country, many of our largest companies ought to be broken up into smaller companies.	—	—	—	—	—
32. The profits of large companies help make things better for everyone who buys their products or services.	—	—	—	—	—
33. Large companies are essential for the nation's growth and expansion.	—	—	—	—	—

The following questions pertain to how much you would like
to see the federal government do in each of the following
areas. Please indicate whether you feel the government should
do a great deal, a fair amount, very little, or nothing.

	A Great Deal	Fair Amount	Very Little	Nothing

1. Owning and operating essential industries.
2. Controlling how much profit a large company can make.
3. Providing medical insurance for doctor and hospital bills.
4. Guaranteeing the prices farmers get for their products.
5. Guaranteeing a job to everyone able to work.
6. Giving financial aid to local and state education.

We would appreciate your answers to a few brief background questions.

1. Are you male or female?   __ female   __ male
2. What is your marital status?
   __ single   __ separated   __ widowed   __ married   __ divorced
3. Do you have any children living at home?   __ yes   __ no
   If yes, please indicate the *number* of children you have at home in the
   following age brackets:
   __ 3 years or younger   __ 13-15 years
   __ 4-5 years   __ 16-19 years
   __ 6-12 years   __ 20 years or older
4. What is your age?
   __ 24 or under   __ 25 to 29 years   __ 30 to 34 years
   __ 35 to 39 years   __ 40 to 44 years   __ 45 to 49 years
   __ 50 to 54 years   __ 55 to 60 years   __ over 60 years
5. What is the highest level of education you completed? Include trade school,
   business school, or other technical training.
   __ Grade school or less
   __ Some high school
   __ Finished high school (graduated)
   Quit high school before graduating and went to trade, technical, or business
   __ school
   __ Graduated from high school and went to trade, technical, or business school
   __ Some college (including junior college)
   __ Finished college (graduated)
   __ Some graduate or professional school
   __ Received a graduate or professional degree
6. a. Are you employed __   unemployed __   retired __ or a student __?
   b. What is your occupation? Please be as specific as you can.
   _____

7. The next questions pertain to your *spouse* if you are married. (If you are single, please go to question number "8")

    a. What is your *spouse's* age?

    —24 or under   —— 45 to 49 years

    —25 to 29 years   —— 50 to 54 years

    —30 to 34 years   —— 55 to 60 years

    —40 to 44 years   —— over 60 years

    b. What is the highest level of education your *spouse* completed?

    —Grade school or less

    —Some high school

    —Finished high school (graduated)

    —Quit high school before graduating and went to trade, technical, or business school

    —Graduated from high school and went to trade, technical, or business school

    —Some college (including junior college)

    —Finished college (graduated)

    —Some graduate or professional school

    —Received a graduate or professional degree

    c. Is your *spouse* employed __ unemployed __ retired __ or a student __?

    If employed, what is his/her occupation? Please be as specific as you can.

    _____

8. Please check the category which best describes your *total family income* for 1974.

    —Under $5,000   —— $15,000 to $19,999

    —$5,000 to $9,999   —— $20,000 to $24,999

    —$10,000 to $14,999   —— $25,000 or above

9. What is your race or ethnic group?

    —White

    —Black

    —Mexican-American

    —Other (please specify) _____

Angell and Associates, Inc. 1975.
A Qualitative Study of Consumer Attitudes Toward Energy Conservation. Chicago: Bee Angell and Associates.

Subject: Conservation behavior, public attitudes concerning energy situation.

Suvey Date: Ongoing from late 1975.

Methods: Series of 10 focus groups of 8-10 people were conducted in four different regions of United States. Groups were moderated by a professional interviewer and followed a semistructured discussion outline. Participants were selected from a widely heterogeneous cross section and were given a cash incentive.

Analysis
Techniques: Interviews were taped for review and analysis.

Significant   The American people are willing to make sacrifices to solve the
Findings: energy problem—only if the need is genuine and responsibility is shared by all.

     Respondents' favorable reaction to this type of study resulted from feelings that the government was sincere in an effort to listen and respond to the needs of the people.

     Most common reactions to the energy issue were characterized by frustration (almost anger) and a sense of helplessness. Also, a consensus that those in a position to exploit the situation are doing just that.

     Most respondents did not think the energy situation was a "crisis"—although they were aware of the impact, they were still able to cope with it. A "crisis" came to mean a drastic curtailment in the availability of energy at a manageable price.

     Responsibility and blame for the current state of affairs were placed on oil companies, public utilities, "business," and the government—not the Arabs or the OPEC countries.

     Respondents felt optimistic about future—especially because of U.S. technological "know-how" and American people's willingness to answer the call.

Since respondents felt the energy situation was not yet critical, they viewed suggestions that large environmental sacrifices were necessary for further energy production somewhat skeptically. If problem does become critical, public seems to be aware that drastic measures may be warranted.

Many were reluctant to conserve because they do not feel "all" are pulling their share of the load; do not wish to give up conveniences; feel exploited and frustrated; and think that big business, industry, and government should do most because they consume the most energy.

Barnaby, David J., and Richard C. Reizenstein. 1975.
Profiling the Energy Consumer: A Discriminant Analysis Approach. Knoxville: University of Tennessee.

Subject: To discover extent of responsiveness to energy crisis (behavioral and attitudinal), to identify consumer segments willing to reduce consumption, and to formulate conservation programs.

Survey Date: February 1974; repeated October 1974.

Methods: Mail questionnaire: 2,500 sent, 922 usable returns (39.4 percent). Random sample from telephone directories: Columbus, Georgia (low pollution); Charlotte, North Carolina (medium pollution); Chattanooga, Tennessee (high pollution) sampled proportionately according to population.

Analysis
Techniques: Multivariate discriminant analysis resulting in profiles of high, medium, low gasoline consumer groups and home heat preference groups.

Significant
Findings:

Profiles of Home Heat Preference Groups	
I	II
Prefer Less Home Heating, with Less Pollution	Prefer Same Amount of Heating and Pollution
Smaller group	larger group
more family members age 15-19	fewer family members age 15-19
more paid family members	fewer paid family members
higher educational level	education not as great

higher income	income not as high
greater use of media	less use of media
and personal	and personal
information sources	information sources
greater willingness to	less willingness to
pay to reduce air	pay to reduce air
pollution	pollution

Major factor in identifying energy-conscious consumer is exposure to media and personal information sources; income also effective discriminator.

Major changes February to October 1974: increased awareness that energy resources are insufficient; greater agreement that resource rationing will be necessary; increased agreement with controlling home temperature by law.

Bartell, Ted. 1974.

"The Effects of the Energy Crisis on Attitudes and Life Styles of Los Angeles Residents." Presented at the 69th Annual Meeting of the American Sociological Association, Montreal (August).

Subject: Behavioral and attitudinal effects of energy crisis and likely impacts on general political orientations and public policies.

Survey Date: February, March 1974.

Methods: Area probability sample of 1,069 Los Angeles County adults, oversampling of blacks to achieve more respondents in the "analytic domain"; interviews.

Analysis
Techniques: Multiple regression.

Significant   20 percent expressed belief in severe energy shortage, 48 percent
Findings: said it was mild, and 26 percent believed no shortage existed; 59 percent said crisis had affected them in some way, but only 6 percent said it had made life much more difficult.

Most reported making efforts to conserve, especially turning out lights when not needed (93 percent) and reducing heating or thermostat setting (80 percent). Only 18 percent reported changing mode of travel to work.

Only significant relationship between conservation efforts and attitudinal or demographic variables was positive relationship between personal conservation and expected future effect on one's own employment.

30 percent blame oil companies; nonbelievers most likely to blame oil companies; blacks and women least likely.

Blaming president related to black ethnicity, lower socio-economic status, female sex.

Energy policies having little or no personal cost generally accepted; 55 mph speed limit (86 percent agree); required 10 percent reduction in use of electricity (75 percent agree); reserved freeway lanes for buses and car pools (70 percent agree).

Bultena, Gordon L. 1976.
Public Response to the Energy Crisis: A Study of Citizens' Attitudes and Adaptive Behaviors. Ames: Iowa State University.

Subject: Attitudinal and behavioral responses of Des Moines, Iowa, residents—especially social-class differences—to high energy prices.

Survey Date: Summer 1974.

Methods: The Des Moines census tracts were ordered on four socio-economic indicators: occupation of residents, educational attainment, house value, average monthly rent. Four census tracts were selected from the total for a high, intermediate, and low position along the above four indicators. Of the 243 eligible respondents (age 18 and over), 190 persons were interviewed. Interviews about one hour long were conducted in the respondents' homes.

Analysis
Techniques: Differences among the three social-class groups were tested for statistical significance using chi-square.

Significant   Most respondents attributed recent energy shortages to large oil
Findings: company actions, government favoritism to the companies, and wasteful energy consumption. Few respondents felt shortages resulted from dwindling energy reserves. Upper class: energy shortages due to dwindling energy supplies, wasteful energy use, and population growth. Middle and lower classes: shortages due to large oil company actions and government favortism to these companies.

Respondents felt energy shortages were less severe locally than elsewhere, and some felt there were beneficial aspects to energy shortages—for instance, increased conservation awareness by the public.

Increased gasoline and home heating costs were the impacts reported most often by respondents.

Lowered thermostats and reduced electric consumption were reported by most respondents, with few other additional conservation actions being practiced (such as car-pooling, reduced leisure travel). More upper-class than middle- and lower-class persons reported taking energy conservation actions.

Few respondents had taken action to influence public decision making on energy policies; upper class most active in this area also.

Greatest priority given to securing sufficient energy for present needs; goals of energy independence, quality of the environment, and low energy prices were important to many respondents, however. Upper class: environmental quality important; lower class: low energy prices important.

Technological "solutions" to energy problems more popular than public policies designed to promote more efficient energy use and/or lessened energy demand. Government largely looked to for amelioration of energy shortages.

Burdge, Rabel J.; Paul D. Warner; and Susan D. Hoffman. 1976. "Public Opinion on Energy." Unpublished paper, University of Kentucky.

Subject: Attitudes toward various conservation measures.

Survey Date: 1976.

Methods: Statewide survey of 3,428 Kentuckians.

Analysis
Techniques: Frequencies.

Significant   Respondents were generally willing to accept voluntary energy-
Findings: saving measures (both transportation and home use).

Strong support for government funding of the development of new sources of energy, but mixed feelings concerning different methods of government regulation of energy use.

Carter, Lewis F., Project Director.* Ongoing.
Interactive Monitoring System for Evaluating Energy Policy Effects on Private Nonindustrial Consumption. Pullman: Washington State University, Social Research Center.

Subject: Establishment of a continuously updated interactive date-retrieval system to monitor consumer energy conservation and the effects of energy shortages and policies.

Survey Date: Ongoing from 1975.

Methods: Statewide monitoring system for policy makers using household consumption data from utility suppliers; initial contact interview with consumer; periodic postcard questionnaires; yearly attitudinal interviews; and contingency questionnaires when households experience changes in employment, residence, or automobile ownership.

Analysis
Techniques: Rotating panel design with 6 panels selected each year. Random sample, stratified by area, used to select 3,100 Washington residents for inclusion in each year's panels. Differences in matched time-lag changes, displacement of time-series data, and perturbations within specific periods will be examined.

Significant
Findings: Not yet reported.

Cook, Stuart W.; Lou McClelland; and Laura Belsten. Ongoing.
Encouraging Energy Conservation in Master-Metered Buildings. Boulder: University of Colorado.

Subject: How can occupants of master-metered office and residential apartment buildings (who do not pay their own energy bills) be encouraged to conserve?

Experiment
Date: September 1976-June 1977.

Methods: Two methods will be contrasted in 4 pairs of University of Colorado office-classroom buildings and in 3 pairs of dormitories:

---

*An asterisk indicates that a survey is particularly significant on the basis of sample, substantive area, analysis techniques, and data comparability.

1. Management, in which the building leader initiates and is responsible for a program of energy conservation by a traditional management approach.
2. User, in which building occupants participate in and carry out all decisions regarding the conservation program.

In a second study, one of a pair of married student apartment complexes will institute a program so that residents whose thermostats are set below a specified level (as ascertained in an unannounced visit from research staff) will be rewarded with a ticket for a weekly lottery drawing. The amount of money in the lottery pool will vary with the amount of energy reduction achieved by the entire complex. A random sample of apartments will be visited each week. The second complex will be a control.

Analysis
Techniques: In both studies, actual use after implementation will be compared with predicted use. Predictions, derived by multiple regression, will be based on past use, weather, amount of sunlight, and other factors affecting use but not under control of the occupants.

Significant
Findings: Not yet available.

Cook, Stuart W.; Lou McClelland; Nancy Wascoe; and Laura Belsten. Ongoing. *A Comparison of Three Methods of Encouraging Homeowners to Install Insulation*. Boulder: University of Colorado.

Subject: What type of persuasive communication, or combination of types, is most effective in encouraging homeowners to install attic insulation?

Experiment
Date: September 1976-February 1977.

Methods: Subjects (firemen who own homes in the Denver metropolitan area):

1. Completed a questionnaire on current patterns of energy use
2. Read one of 7 communications, or no communication
3. Completed a questionnaire on opinions, attitudes, and behavioral intentions regarding energy use

4. (1 month later) were invited to have a free inspection of home insulation needs by the Public Service Company of Colorado.

The 7 communications were the following:

1. An "energy crisis" appeal—there will be an energy shortage, it will have adverse effects on our lives, but we can help avert it by conservation
2. An economic appeal outlining the monetary savings resulting from installation of insulation
3. "How-to" booklet explaining how to install insulation or have a contractor do it
4. Energy and economic
5. Energy and how-to
6. Economic and how-to
7. Energy, economic, and how-to.

Analysis
Techniques: 2 x 2 x 2 analyses of variance or chi-square analyses on attitudes and behavioral intentions, acceptance of insulation inspection, and actual installation of insulation

Significant
Findings: Not yet available.

Curtin, Richard T. 1975.
             "Consumer Adaptation to Energy Shortages." Ann Arbor: University of Michigan, Survey Research Center (unpublished manuscript).

Subject: Conservation behavior, attitudes, and motivations.

Survey Date: Autumn 1974.

Methods: Personal interviews with family heads or spouses from 1,400 randomly selected family units within coterminous United States.

Analysis
Techniques: Multiple-classification analysis.

Significant   Widespread past conservation efforts reported; widespread
Findings: prospect of difficulty in making future adjustments found among all subgroups and socioeconomic levels.

Personal experience with past conservation significantly lowers expectation of difficulty of future adjustments to energy shortages.

Those who believe government can effectively handle national economic problems have made greater conservation efforts and view future efforts as less difficult.

Young and highly educated report greater conservation efforts in past, see less future difficulty in conserving heat and electricity; no significant relationship to income.

Smaller homes—less past conservation.

Larger homes—future conservation viewed as more difficult.

Family size—curvilinear relationship to conservation.

Dwellers in large metropolitan areas report easy adjustment to gasoline conservation, but difficulty in reducing home heating; rural dwellers report the opposite.

Doering, Otto C.; Jerry Fezi; Dave Gauker; Mike Michaud; and Steve Pell. 1974. *Indiana's Views on the Energy Crisis.* Lafayette, Indiana: Purdue University, Agricultural Economics Department, Cooperative Extension Service.

Subject: Opinions regarding energy crisis.

Survey Date: April, May 1974.

Method: Mail questionnaire sent to 1,000 Indiana residents, drawn randomly from automobile registration lists (670 returns).

Analysis
Techniques: Frequencies.

Significant   50 percent did not really believe there was an energy crisis. Oil
Findings: companies were blamed most often (34 percent) for energy problems and were accused of profiteering, holding back production, and hoarding fuel.

66 percent thought the federal government had not done a good job in dealing with the energy crisis, but only 33 percent thought the state government had not done a good job.

36 percent indicated that the energy crisis had had a real effect on the way they lived.

Majority had made some effort to reduce home heating and gasoline consumption.

Least difficult and costly conservation policies were most strongly favored, even at expense of increasing pollution to some degree.

Doner, W.B., Inc., and Market Opinion Research. 1975.
>*Consumer Study–Energy Crisis Attitudes and Awareness.*
>Lansing: Michigan Department of Commerce.

Subject: Awareness, attitudes, behavioral change, and perceived future effects of the energy crisis; changes over time.

Survey Date: February 1974, 1975.

Methods: Telephone interviews with stratified area sample of 400 (1974) and 525 (1975) residents of Michigan.

Analysis
Techniques: Frequencies.

Significant Virtually all Michigan adults are aware of energy crisis and con-
Findings: servation publicity, but only 50 percent perceive a real energy crisis. This perception is up 9 percent since the 1974 survey.

>75 percent claim to have changed their behavior because of the crisis. People believe that these changes will be permanent and that the energy problem will be a long-term one (5 years).

>Major motivator for conservation measures is that "conserving energy saves money."

>Those who believe there is an energy crisis behave in a more conservation-minded fashion than those who are cynical and distrustful of the reports of such a crisis.

Donnermeyer, Joseph F. 1976.
>"Social Status and Attitudinal Predictors of Intentions Toward Practicing Energy Conservation Measures and Energy Consumption Behavior." Unpublished Ph.D. dissertation, University of Kentucky.

Subject: Consistency among attitudes, intentions, and behavior regarding energy conservation.

Survey Date: Spring 1976.

Methods: Statewide survey of Kentucky citizens' attitudes and opinions. Consumption data collected for Fayette County subsample.

Analysis
Techniques: Simple and partial correlation techniques.

Significant
Findings: Not yet available.

Eichenberger, Mary Ann. 1975.
  "A Comparison of Ownership of Selected Household Appliances and Residential Energy Use by Employed and Nonemployed Homemakers in the Lansing, Michigan, Area." Unpublished master's thesis, Michigan State University.

Subject: Differences in residential energy use between households with employed and nonemployed homemakers.

Survey Date: May, June 1974.

Methods: Multistage area probability sample (interviews and self-administered questionnaires) of 187 families in Lansing, Michigan, SMSA.

Analysis
Techniques: Analysis of covariance.

Significant   No significant differences were found among full-time, part-
Findings: time, and nonemployed on type or quantity of appliances owned.
              Although differences were not statistically significant, a trend emerged indicating that nonemployed homemakers use the greatest amount of direct residential energy, and full-time employed use the least.

Gallup Organization, Inc. 1976.
  Group Discussions Regarding Consumer Energy Conservation. Washington, D.C.: Federal Energy Administration.

Subject: Consumer attitudes toward the energy crisis and conservation communications.

Survey Date: Early 1976.

Methods: 8 group discussions of 8-10 participants (4 in Trenton, New Jersey, 4 in Denver, Colorado) with participants from various consumer segments (age, sex, home ownership, urban versus suburban residence, past conservation behavior), led by experienced moderators.

Analysis
Techniques: Taped for review and analysis.

Significant   Skepticism, cynicism, ignorance dominated all groups, as did a
Findings: general suspicion of being constantly lied to regarding energy.

For most people, saving energy is simply a process of ritualistic acts rather than realistic efforts. It is accurate to think of consumers on a continuum from relatively low wasters to relatively high wasters. Most saving is thought of in terms of monetary frugality.

Participants hear "deny yourself" as the implicit theme in most conservation communications and are answering with "I have earned the right to indulge." Consensus was that we are a spoiled, self-indulgent nation unconcerned with the future; and individuals are reluctant to take responsibility, abdicating it to the government.

Convenience and immediate gratification are primary goals, limited only by financial pressure. Saving energy when it is "convenient" provides sense of contributing and helps relieve guilt.

Self-indulgent consumption values displayed in most advertising are conflicting factor.

Young, middle-class adults show more concern with quality of life than material success, and incorporate conservation ethic into larger value system stressing simplicity, low technology, and conservation in the largest sense, but do not see energy conservation as being particularly constructuve without massive realignment of social priorities. This group is skeptical and/or cynical as to the ability of the private sector to resolve or even comprehend the energy problem. Desire to save money is still the most effective motive.

Attitudes toward alternative sources—Energy conservation viewed as a temporary, time-buying strategy. Alternative sources are expected to make it possible to continue as a spoiled, indulgent nation.

General agreement that monetary incentives are the most effective motivators. Amount of money saved is only meaningful measure of energy to those financially motivated. "Dollar saving" strategy may affect conservation behavior, but will not change attitudes.

Pre-teens and energy conservation—"Energy crisis" remote and confusing. Most feel adults have major responsibility for conservation. Adults are perceived as self-indulgent wasters; but this model is not rejected, as most discussants are already socialized into the "American Dream." React to possibility of resource depletion by expecting either a catastrophic return to primitive time or a scientific/technological breakthrough "just in time."

Teenagers—Aware of problem, but not personally affected by it. Some anger expressed at world for bequeathing this problem to them. Most feel the major responsibility belongs to adults, and respond only to price.

Gladhart, Peter Michael. 1976.
Energy Conservation and Lifestyles: An Integrative Approach to Family Decision-Making. Occasional Paper no. 6, Family Energy Project. East Lansing: Michigan State University, College of Human Ecology, Institute for Family and Child Study.

Subject: Role of energy is supporting the family ecosystem, energy costs of different lifestyle choices.

Survey Date: Spring 1974.

Methods: Use of self-administered questionnaires and personal interviews with random sample of families in the Lansing area. Consumption data secured from utility companies and fuel dealers.

Analysis
Techniques: Frequencies, crosstables, average energy consumption.

Significant   Single-family homes require much more energy than either
Findings: multifamily housing or mobile homes.
                      Life-cycle stage of family and family size influence energy use. Families containing no children or families where the wife is at least 60 years old use about 13 percent less energy.
                      Average energy cost per person in family of 5 or more persons is less than 50 percent of the per-person cost in a 2-person family.
                      No important differences between urban and rural residential consumption, but rural residents use substantially more gasoline.

Gollin, Albert E.; Shirley J. Smith; and Joanne S. Youtie. 1976.
New Hampshire Energy Usage Patterns and Consumer Orientations: A Comparative Assessment. Washington, D.C.: Bureau of Social Science Research, Inc.

Subject: Public awareness of and attitudes toward energy issues.

Survey Date: April 1976.

Methods: Random sample drawn from telephone directories in New Hampshire. Telephone interviews conducted.

Analysis   Frequencies, cross tabulations. Index of acceptance scores for
Techniques: time-of-day pricing.

Significant   Almost 66 percent were "very concerned" about the amount
Findings: of electricity used in their homes. Those in family-forming age
                groups, those with lower income levels, homeowners, and those
                with higher monthly bills were most likely to express concern.

                Awareness of kwh used and knowledge about billing pro-
                cedures and meter reading were generally low, but were higher
                among those who expressed concern over the amount of elec-
                tricity used in their homes.

                Time-of-day pricing not familiar to most respondents.
                Resistance to changes in the timing of routine activities was
                substantial. Laundry and dishwashing were the most flexible,
                cooking and bathing the least.

                About 20 percent of the sample scored high in their accept-
                ance of time-of-day pricing, 50 percent scored low. Size of the
                monthly bill, expressed concern, and family income are signifi-
                cantly and positively correlated with acceptance score.

Gottlieb, David. 1974.
                Sociological Dimensions of the Energy Crisis. Project E/S-5.
                Austin: State of Texas, Governor's Energy Advisory Council.

Subject: Perceptions, attitudes, behavior, and expectations of Texans in
                response to energy crisis.

Survey Date: April May 1974 (preembargo, southern Texas); June, July 1974
                (postembargo; northern Texas).

Methods: Sample of heads of households living in urban (Houston, Amarillo)
                and rural (Colorado County, Deaf Smith County) areas of Texas;
                urban—random sample of year-round housing units from census
                block data types; rural-random sample of names and addresses
                from county tax rolls; mobile and poor underrepresented; ques-
                tionnaires hand-delivered and retrieved—845 + 250 replacements,
                782 returned.

Analysis
Techniques: Frequencies, cross tabulations, chi-square.

Significant   Only major difference found between two regional samples was
Findings: a greater concern about anticipated escalating costs of energy
                expressed by the postembargo (northern Texas) sample.

                Both samples (preembargo and postembargo) show these
                similarities:

1. Fail to see energy situation as reflective of a long-term serious crisis

2. Show distrust of producers and distributors of energy, along with government officials responsible for energy policies and programs

3. Feel citizens are energy-wasteful; environmentalists are not to blame

4. Lack of knowledge of energy origins and amount of energy consumed by appliances and devices correlated with lack of belief in crisis.

Socioeconomic status of citizens is important to energy behavior and attitudes: poorer people have fewest alternatives and therefore are affected most by energy situation.

Data show that "The more real the perception of the crisis or emergency, the more responsible the populace will be." Major reasons cited for energy shortages:

1. Desire for bigger profits by oil and utility companies

2. Wastefulness of citizens

3. Arab oil embargo.

"More citizens believe that the 'entire shortage is part of a political scheme' than agree that 'the world is running out of fuel.'"

Data show majority of respondents will believe in an energy crisis when:

1. They cannot buy or have available the energy necessary for their personal needs

2. They "see" all citizens are equally deprived of available energy sources

3. Proposed policies match expressed severity of energy crisis.

Although a strong consensus was expressed that citizens have been wasteful and conservation should be practiced, the levels of voluntary conservation that people practice or are likely to practice appear quite minimal.

Authors conclude that what is needed is real evidence of an energy crisis; a systematic and well-conceived education program for the public; an equitable and reasonable system of energy conservation incentives; energy conservation policies that stress economic benefits instead of energy shortages.

Gottlieb, David, and Marc Matre.* 1976.
Sociological Dimensions of the Energy Crisis: A Follow-up Study.
Houston, Texas: University of Houston, Energy Institute.

Subject: To determine extent of change in energy conservation behavior, attitudes, and values in Texas since previous 1974 study.

Survey Date: April-June 1975.

Methods: Same names and addresses as in first survey, plus replacements for lost cases and inclusion of newly built-up areas (random-sampled from newly constructed streets); questionnaires hand-delivered—938 retrieved (89 percent).

Analysis Techniques: Those answering in both 1974 and 1975, regardless of place of residence in 1975, used in panel analysis. Those sampled in 1975 only were used in trend analysis. Scales constructed: energy knowledge, environmental concern; frequencies, cross tabulations, measures of association, and significance.

Significant Findings: Majority accept propositions that world is running out of fuel and that Americans are wasteful, but only a slight increase in belief in serious, long-term energy crisis. Seen as just another problem, and not as important as the social problems, such as inflation, crime, unemployment. Main motivation for those who conserve is cost. Therefore, while higher socioeconomic persons are more likely to believe in crisis, lower- and middle-status people are more likely to reduce energy usage.

Majority not knowledgeable about energy and conservation matters.

Willingness to endorse policies that will cause least inconvenience or disturbance of life style.

Oil companies bear brunt of public resentment and blame for the energy situation.

Low energy knowledge and disbelief in an energy crisis are correlated, and both are proportionately more frequent among females and persons of low socioeconomic status. Of respondents who believe in an energy crisis, 68 percent report a "changed way of living," compared with 48 percent of those who do not agree that there is an energy crisis. The most widespread conservation area involves savings on utilities, followed by reduction in short-distance driving. In terms of worrisome consequences of the energy situation, cost is inversely related to age and socioeconomic status; shortages are directly related.

On causes of the energy problem, the following patterns emerge:

1. No belief in crisis (generally younger, lower socioeconomic status). Scheming between oil companies; government

attempts to draw attention from real problems; United States has exported too much fuel; technology has not kept pace with needs.

2. Belief in crisis (generally older; higher socioeconomic status). Consumers shift away from use of coal; failure to build nuclear power plants; world is running out of fuel; efforts of environmentalists.

Grier, Eunice S. 1976.

"Changing Patterns of Energy Consumption and Costs in U.S. Households." Paper presented at Allied Social Science Associations meeting, Atlantic City, New Jersey (September).

Subject: Energy awareness, attitudes, and consumption behavior.

Survey Date: Spring 1973; 1975.

Methods: Two representative national samples of households surveyed by Washington Center for Metropolitan Studies: in-depth interviews, direct measurement of utility consumption and costs. Post-crisis sample numbered 3,200; precrisis sample was half that size.

Analysis
Techniques: Frequencies, cross tabulations, average consumption.

Significant   Energy costs, while rising, remain a relatively small part of the
Findings: average household budget. Nevertheless, for certain categories of people (notably the poor and the elderly), rising energy costs are a serious and growing burden.

Most Americans are aware that the United States has an "energy problem," and many say they are doing something in their personal lives to deal with it. Efforts usually are limited, however, to those things that are easiest to do.

Only a small minority of households had made substantial energy-saving improvements to their dwelling unit. Middle-income groups ($14,000-16,999) made the largest proportion of energy-conserving improvements.

Energy conservation measures generally were directly and positively related to income, with some exceptions in middle-income categories.

More than 50 percent of the number surveyed definitely agreed that every family should be willing voluntarily to reduce energy consumption to no more than the average amount needed by a family of the same size.

Hass, Jane W.; Gerrold S. Bagley; and Ronald W. Rogers. 1975.
"Coping with the Energy Crisis: Effects of Fear Appeals upon Attitudes Toward Energy Consumption." *Journal of Applied Psychology* 60: 754-56.

Subject: Experiment to test two components of communication messages upon intentions to reduce energy consumption.

Experiment
Date: 1975.

Methods: 2 x 2 factorial design manipulating reported noxiousness of an energy crisis and probability of occurrence; 60 undergraduate students.

Analysis
Techniques: Analysis of variance for main and interaction effects.

Significant   Consonant with other findings regarding communication variables
Findings: that are efficacious in producing attitudinal change, the single best predictor is the reported noxiousness of an event. Authors conclude that magnitude of potential threat is more conducive to changed attitudes than the reported probability of occurrence.

Heberlein, Thomas A. 1975.
"Conservation Information: The Energy Crisis and Electricity Consumption in an Apartment Complex." *Energy Systems and Policy* 1:105-17.

Subject: Effect on consumption behavior of informational material designed to increase or decrease electricity use.

Experiment
Date: March, April 1973, 1974.

Methods: Readings taken at 6 pm of electric meters at 84 apartments in 6 buildings near Madison, Wisconsin; apartments rented mostly to young couples and families for $145-175/mo.
                Letters designed to manipulate variables (electricity cost beliefs, beliefs of consequences to others, and consumer responsibility) sent, follow-up phone call urging reading of letter; 1 control group, 3 experimental (2 treated to decrease consumption, 1 to increase).

1974—meters reread during same months to assess impact of year of media information; 73 of original 84 apartments occupied; 43 by same families.

Analysis
Techniques:
Measurement of consumption over time period; mean daily use, standard deviation.

Significant
Findings:
informational material had no effect on consumption. No changes in electricity consumption during energy crisis. Wide variability of consumption within group studies (similar social class).

Hogan, Mary Janice. 1976.
"Energy Conservation: Family Values, Household Practices, and Contextual Variables." Unpublished Ph.D. dissertation, Michigan State University.

Subject: Family values and conservation behavior.

Survey Date: May, June 1974.

Methods: Multistage probability sample (interviews, self-administered questionnaires) of 157 families in Lansing, Michigan, SMSA.

Analysis
Techniques:
Value scales established (self-esteem, familism, social responsiveness, and ecoconsciousness) and tested as predictors of conservation behavior.

Significant
Findings:
No systematic relationship was found between conservation behavior and demographic characteristics.
Ecoconsciousness (awareness of finiteness of energy-resources and wastefulness of certain consumption patterns) positively related to adoption of conservation measures.
Social responsiveness related to various demographic features, but not to conservation behavior.
Familism, self-esteem related to neither demographics nor conservation behavior.

Honnold, Julie A., and L.D. Nelson. 1976.
"Voluntary Rationing of Scarce Resources: Some Implications of an Experimental Study." Paper presented at Annual Meeting of the American Sociological Association, New York (August).

Subject: Scarcity information and conservation behavior.

Experiment
Date: 1976.

Methods: Sample consisting of 485 undergraduate students completed
            questionnaire. Subsample of 218 viewed scarcity information film.

Analysis   Scales developed for perceived necessity of conservation) posttest
Techniques: and pretest), perceived sufficiency of conservation (posttest and
            pretest) commitment to conservation behavior, avoidance of
            scarcity information.

Significant Commitment to conservation positively and significantly related
Findings: to perceived necessity of conservation.
                    Perceived sufficiency and conservation commitment signifi-
            cantly and positively related.
                    The more remote the perception of social consequences of
            resource depletion, the lower the commitment to conservation
            behaviors.
                    Conservationists and cynics have higher commitment to
            conservation behavior than do consumerists. Consumerists tend
            to more persistently avoid contact with scarcity information.
                    Those who see the effects of resource depletion as remote
            have higher avoidance scale scores than those who consider the
            problem more immediate.
                    Consumerists were changed the least by the filmviewing.
            33 percent of the conservationists became cynics or consumerists.

Hummel, Carl F.; Lynn Levitt; and Ross J. Loomis. 1975.
        *Perceptions of the Energy Crisis: Who Is Blamed and How Do
        Citizens React to Environment-Lifestyle Tradeoffs?* Fort Collins,
        Colorado: Colorado State University, Department of Psychology.

Subject: Effect of energy crisis on support for mandatory and voluntary
            energy and pollution programs.

Survey Date: Late summer, November 1973. One sample surveyed when gaso-
            line was suddenly scarce ("acute") and one when energy problem
            was well-established ("chronic").

Methods: Mail questionnaires to representative sample of residents of Fort
            Collins, Colorado. "Acute" sample n = 114, "chronic" sample
            n = 124.

Analysis
Techniques: Stepwise regression.

Significant   Most demographic and perceived-personal-effect variables had no
Findings: consistent, significant effect on support for voluntary and manda-
               tory programs.
                   In both samples, blaming environmentalists for the energy
               crisis was negatively related to support for mandatory programs
               that would attack both air pollution and energy problems.
                   Blaming environmentalists was a positive predictor for
               proenergy actions that would damage the environment.
                   Those blaming individual consumers were more likely to
               support mandatory remedies for energy and pollution problems.

Hyland, Stanley E.; Judith S. Liebman; Demitri B. Shimkin; Richard C.
Roistacher; James J. Stukel; and John J. Desmond. 1975.
        *The East Urbana Energy Study, 1972-1974: Instrument Develop-
        ment, Methodological Assessment, and Base Data.* Champaign,
        Illinois: University of Illinois, College of Engineering.

Subject: Change in behavior and attitude regarding energy and conservation.

Survey Date: Fall 1972, Spring 1973; follow-up in June 1974.

Methods: 10 percent stratified random sample of households in East
               Urbana, Illinois; questionnaire administered in personal inter-
               views.

Analysis
Techniques: Frequencies; degrees of association.

Significant   (Major findings not yet published.) Population responded to
Findings: energy crisis, due to rising costs, by using air conditioners,
               vacuum cleaners, and ovens less. High value emphasis on privacy,
               autonomy, and mobility expressed in automobile ownership and
               use, and little change in use patterns was found. Authors con-
               clude that conservation strategies will not be effective if they
               affect "deeper life-styles," meaning the value systems of a con-
               sumer.

Kilkeary, Rovena. 1975.
        "The Energy Crisis and Decision-Making in the Family." Spring-
        field, Va.: National Technical Information Service, U.S. Depart-
        ment of Commerce. PB238783.

Subject: Whether family characteristics and energy-related experiences
               affect household energy knowledge and conservation practices.

Survey Date: July-August 1974.

Methods: Interviews with 602 randomly selected households in Bronx and
                Queens, New York. (Queens area affected by extended blackout
                preceding summer; Bronx not affected.)

Analysis   Energy knowledge and changed practices scales, correlations,
Techniques: analysis of variance.

Significant   Related positively to energy knowledge scores—car ownership,
Findings: education, family composition.
                    Related positively to changed practices scores—exposure to
                extended blackouts, direct payment of utility bills, car ownership,
                belief in family effort to produce change affecting the energy
                crisis, family composition.
                    Strongest influence on knowledge and conservation is
                income (highest scores, middle income). Families composed of
                couples with children demonstrate high levels of energy know-
                ledge and conservation practices.

Morrison, Bonnie Maas. 1975.
                "Socio-physical Factors Affecting Energy Consumption in Single
                Family Dwellings: An Empirical Test of a Human Ecosystems
                Model." Unpublished Ph.D. dissertation, Michigan State University.

Subject: Influence of structural housing conditions and socioeconomic life
                style on energy consumption and attitudes.

Survey Date: May, June 1974.

Methods: Multistage area probability sampling in selected urban and rural
                areas of a mid-Michigan SMSA; energy consumption data for each
                household from utility company and fuel oil sources (n = 217).
                First stage of a five-year longitudinal study.

Analysis
Techniques: Stepwise regression and path analysis.

Significant   Physical housing factors (number of persons, number of appli-
Findings: ances, number of rooms) were more highly correlated with energy
                consumption than were life-style factors.
                    Belief in reality of the energy problem is positively related
                to mean (husband-wife) educational attainment, agreement
                between husband and wife on beliefs regarding availability of

electrical energy, and reported total costs of all energy forms used in the household.

Belief in reality of crisis not found to effect change in consumption patterns.

Morrison, Bonnie Maas, and Peter M. Gladhart.* 1976.
"Energy and Families: The Crisis and the Response." *Journal of Home Economics* (January): 15-18.

Subject: Family life style, attitudes, and consumption practices.

Survey Date: Spring 1974.

Methods: Random survey of residents in Lansing, Michigan, metropolitan area. Consumption data also available. Initial stage of five-year longitudinal study.

Analysis
Techniques: Frequencies, average energy consumption.

Significant   Family income was single best predictor of residential energy
Findings: consumption; richer families consume more than poorer families.

Families in child-rearing stages use more energy than families without children or those at later stages of the life cycle.

Employed homemakers use less energy than nonemployed, yet own the same number and types of appliances.

Housing relates directly to energy consumption (size, single or multifamily arrangement, number of rooms, doors, and windows), but these variables are closely related to family income.

Families studies were evenly divided on belief in the reality of the energy crisis. Belief, in itself, does not decrease consumption.

Families where both husband and wife were conscious of the finiteness of energy resources and the wastefulness of consumption patterns were more likely to have adopted conservation practices than those families where consciousness was lower or where husband and wife differed in their commitment.

Urban and rural families differed sharply in acceptance of various possible energy policies. Rural residents were less accepting than urban residents of tax deductions for apartment dwellers and for small-car owners, gas rationing, single-car deductions, and free mass transit.

Slightly more than 25 percent of the sample found a tax deduction for sterilization or for small families acceptable.

In general, urban women were most favorable to all policies and, in descending order, urban males, rural females, and rural males.

Muchinsky, Paul M. 1976.
"Attitudes of Petroleum Company Executives and College Students Toward Various Aspects of the 'Energy Crisis.'" *Journal of Social Psychology* 98: 293-94.

Subject: Attitudes toward various aspects of the energy crisis.

Survey Date: Spring 1974.

Methods: 26 members of Independent Connecticut Petroleum Association and 328 undergraduate Iowa State University students sampled by questionnaire.

Analysis
Techniques: One-way analysis of variance, rank-order correlation.

Significant   Students perceived major oil companies as being primarily respon-
Findings: sible for the energy crisis by withholding petroleum supplies in order to increase profits and reduce competition.

Petroleum company executives tended to perceive federal government as being primarily responsible by "handcuffing" petroleum industry with taxation, pricing, and import legislation.

Both groups ranked economics, corruption in government, and the energy crisis, as first, second, and third most important social problems, respectively.

Murray, James; Michael J. Minor; Norman B. Bradburn; Robert F. Cotterman; Martin Frankel; and Alan E. Pisarski.* 1974.
"Evolution of Public Response to the Energy Crisis." *Science* 184:257-63.

Subject: Public exposure and reaction to energy crisis—attitudes and behavior.

Survey Date: Weekly since April 1973.

Methods: Small, weekly nationwide probability sample of households; surveys conducted by NOTC—University of Chicago; adults, 18 and older, in 48 contiguous United States; interviews.

Analysis
Techniques: Frequencies, percent distribution, cross-tabulations.

Significant   Majority believe energy shortage is important problem, but only
Findings: 25 percent say it is the most important problem facing nation
                 today.
                        Widespread agreement that federal government and oil
                 companies are responsible for crisis.
                        Large majority say crisis has changed their way of life
                 somewhat, but few perceive this change as major. Little indica-
                 tion of serious change in life style.
                        Majority believe we will have as much energy as we need in
                 5 years.
                        Those reporting difficulty in obtaining fuel are twice as
                 likely to expect further problems over next year.
                        67 percent agree that gasoline shortages could be solved if
                 individual consumers cut down consumption, 78 percent in rural
                 areas.
                        Opinions concerning importance of energy problem unre-
                 lated to education, income, or area of residence; positively
                 related to expectations of problems in obtaining fuel and to
                 reports of change in life style as a result of energy shortages.
                        Evaluations of responsibility unrelated to other variables.
                 Little relationship between conservation behavior and other
                 variables. Exception: shutting off lights related to reports of
                 difficulty in obtaining electricity and belief in importance of
                 energy problem.

Newman, Dorothy K., and Dawn Day.* 1975.
        *The American Energy Consumer. A Report to the Energy Policy
        Project of the Ford Foundation.* Cambridge, Massachusetts:
        Ballinger.

Subject: Household energy use.

Survey Date: Interviews—May, June 1973; billing data—June-September 1973.

Methods: Personal interviews with heads of household—1,455 respondents
                 (65 percent); multistage area probability sample (national),
                 with oversampling of lowest socioeconomic quartile; separate
                 survey of electric and gas companies serving households billed
                 directly, to obtain billing data for those who gave permission
                 (90 percent response rate).

Analysis   Weight factors inversely proportionate to probability of inclusion
Techniques: of each household in sample; descriptive statistics.

Significant   Higher income groups consume more energy (household and
Findings: automobile) regardless of other factors.
                       Basic features of structure of home (over which consumer
                  has little control) more important than choice of appliances in
                  energy consumption. This lack of choice affects blacks and lower
                  income groups to greater extent.
                       Poor use less energy, pay more for it, and are more exposed
                  to noxious by-products of energy consumption and production
                  than other income groups. Of all income groups blacks consume
                  least.
                       Authors conclude that households by themselves can play
                  only a minor role in conservation.

Opinion Research Corporation, Michael Rappeport and Patricia Labaw, Project
Directors.* 1974-75.
                  Public Opinion Polls on Energy. Highlight Reports (see complete
                  list in "References" section of Chapter 2). Princeton, New Jersey:
                  Opinion Research Corporation (for the Federal Energy Adminis-
                  tration).

Subject: Energy-related attitudes and behavior.

Survey Date: Monthly for 20 months, beginning September 1974.

Methods: Telephone interviews; randomly selected adults in households
                  having telephones, nationwide; 600-1,200 interviews/month.

Analysis   Frequencies, cross tabulations, multiple regression on attitudes
Techniques: toward mass transit.

Significant   After one year of study, public remains split about seriousness
Findings: of energy problem; but there has been a slight increase in percent
                  who believe it is serious.
                       No demographic variables correlate with belief in reality
                  of crisis.
                       No significant differences in attitudes between low-income
                  and total population. Behavioral differences are slight and result
                  from structural (especially economic) influences rather than from
                  conscious energy-related decisions.

Energy shortage ranked well below rising unemployment and inflation, as national problem. Little difference among age groups or over time.

Blame of oil companies for crisis diminished over study year, while blame of the wastefulness of American consumers increased.

Dissatisfaction with efforts of President Ford and Congress to alleviate problem increased.

Pervasive lack of knowledge about elementary energy questions not affected by age or income. Slight differences by sex—women more knowledgeable in areas that impinge on daily life, such as price increases; men more knowledgeable about existence and nature of FEA.

Awareness of existence of FEA dropped slightly over time. In 1975, 66 percent not aware it exists.

Effects of energy shortages remained about the same. Most consumers have been affected at least a little. In first waves of interviews, experience with shortages was largely limited to gasoline; but over the year experience spread to other energy sources.

People generally not able to cite accurately amount spent on home fuel.

Consumer groups, federal government, and news media rank high as reliable information sources; business ranks considerably lower.

Household conservation (lighting, appliances, heating, and cooling) efforts increased.

Most respondents believe they are making as much as or more effort than others to conserve.

Importance of price as increasing factor in conservation.

When respondents were asked why people do not try to conserve, most common responses were lack of motivation and lack of belief in existence of an energy crisis.

Possibility of rebates from federal government of 25-50 percent of the cost of insulation and other energy savers appears encouraging.

Public generally does not favor removing pollution controls. Younger persons more willing to pay more for environmental protection.

Downward change over study year in overall favorability regarding nuclear power plants.

Most Americans did not follow federal government advice on thermostat settings during winter of 1974-75, nor do they intend to do so during winter of 1975-76. Reason: Federal government not seen as a good source of advice on energy-saving procedures.

Many people (42 percent) believe personal conservation efforts can have great impact on total energy consumption.

Peck, A.E., and O.C. Doering. 1975.
    Voluntarism and Price Response: Consumer Reaction to the Energy Shortage. Lafayette, Indiana: Purdue University, Department of Agricultural Economics.

Subject: Changes in efficiency of household use of two heating fuels.

Survey Date: 1971-74.

Methods: Sample of 174 natural gas and 279 liquid petroleum customers, all rural residential accounts. Fuel use data acquired.

Analysis
Techniques: Calculation of heating degree-days and measures of average efficiency.

Significant
Findings: Liquefied petroleum customers, whose fuel costs rose considerably, increased fuel use efficiency 14.4 percent from 1973 to 1974.

    Natural gas customers, whose costs rose insignificantly, increased their efficiency 5.8 percent from 1973 to 1974.

Perlman, Robert, and Roland L. Warren.* 1975a.
    Energy Saving by Households in Three Metropolitan Areas. Waltham, Massachusetts: Brandeis University, Heller Graduate School for Advanced Studies in Social Welfare.

Subject: Impact of energy problems on households in metropolitan areas in different regions of the country.

Survey Date: November 1974.

Methods: Towns selected because of region, climate, and primary source of fuel (Hartford, Connecticut; Mobile, Alabama; Salem, Oregon); households selected—multistage probability-sample personal interviews; preference order: female head, male head, other adult; 1,913 contacts, 1,440 completed and processed.

Analysis
Techniques: Frequencies, cross tabulations.

Significant   Energy conservation behavior and attitudes show more similari-
Findings: ties than differences among the three metropolitan areas.

    50-75 percent reported cutting use by driving less, lowering thermostats, and cutting back on electricity.

    Hartford, Connecticut (Northeast, dependent on imported oil), reported greatest conservation efforts.

    62 percent did not believe crisis was real, but contrived to boost oil and gas company profits.

    Nonbelievers—same in terms of saving energy as those who believe crisis was real.

    Price cited as most important reason for saving by 35 percent of motorists and 41 percent of householders who made conservation efforts.

———. 1975b.

    Energy Saving by Households of Different Incomes in Three Metropolitan Areas. Waltham, Massachusetts: Brandeis University, Heller Graduate School for Advanced Studies in Social Welfare.

Survey Date: Same as Perlman and Warren (1975a).

Subject: Impact of energy problems on households of different income groups in three metropolitan areas.

Method: Same as Perlman and Warren (1975a).

Analysis
Techniques: Same as Perlman and Warren (1975a).

Significant   Efforts to conserve vary more according to community than to
Findings: income of household. More conservation in higher income groups.

    Majority report price as reason for conservation—little variance among income groups.

    Appliances reduced most—air conditioners, clothes dryers, dishwashers, fans—considered conveniences. Those reduced least—heaters, stoves, washers—considered necessities.

    Reductions highest in areas where rates are highest.

    Number of "believers" increases as income rises, as does blaming oil and gas companies.

    Lower income—more skeptical about reality and greatest blame on federal government.

Schwartz, T.P., and Donna Schwartz-Barcott. 1974.
    "The Short End of the Shortage: On the Self-Reported Impact of

the Energy Shortage on the Socially Disadvantaged." Paper presented at the 1974 meeting of the Society for the Study of Social Problems. Montreal, Quebec.

Subject: Effect of energy shortages on the socially disadvantaged.

Survey Dates: July 1973; November 1973; March 1974.

Methods: Telephone interviews with a systematic sample (n = 200), proportionate to city size, of heads of households with listed telephone numbers in 3 North Carolina cities.

Analysis   Contingency analysis of cross-tabulated data, chi-square, and
Techniques: various measures of association.

Significant   Respondents generally reported that the energy shortage had
Findings: little or no impact on them or their households, with little difference between the socially advantaged and the disadvantaged. In fact, the advantaged reported slightly more impacts.

                 Impact of the shortage increased for all groups over time, with more reported increase in the advantaged groups than in disadvantaged groups.

                 Groups with multiple disadvantages generally did not differ from the sample as a whole, but there was some evidence of an inverse relationship between the number of social disadvantages possessed by a group and the frequency with which energy shortage impacts were reported.

Sears, David O.; Tom R. Tyler; Jack Citrin; and Donald R. Kinder. 1976.
        "Political System Support and Public Response to the 1974 Energy Crisis." Paper presented at Conference on Political Alienation and Political Support, Palo Alto, California (May).

Subject: Effect of general sentiments about ongoing political order on responses to the energy crisis.

Survey Date: February-March 1974, late 1975.

Methods: Multistage probability sample of Los Angeles County residents, 1,069 age 18 or over and 195 age 12-17. Reinterviews of adult sample in 1975, with permission obtained to use consumption records from utility companies.

Analysis   Frequencies, cross tabulations, gamma statistics, regression
Techniques: analysis.

Significant   Situational pressures, such as personal impact of the crisis and
Findings: local conservation sanctions, have more effect on consumption-
               reducing behavior than do political attitudes such as system
               support and partisanship. Moreover, situational and attitudinal
               variables have no interactive effects on behavior.
                   Diffuse system support was significantly related to accept-
               ance of official government energy line on reality of the crisis,
               support for energy policies, and behavioral compliance on conser-
               vation measures. Partisanship (especially evaluation of President
               Nixon) was also related.
                   Personal impact of the crisis had no effect on the individ-
               ual's attitudinal response to it, and attitudes had no effect on
               behavior.

Seattle City Light. 1976.
               Energy 1990 Study Initial Report. Seattle, Washington: Office of
               Environmental Affairs.

Subject: Alternative options for energy and electricity generation for
               Seattle in the future.

Survey Date: 1975-76.

Method: Questionnaires included in all customers' bills.

Analysis
Techniques: Report on questionnaire survey forthcoming in *Final Report.*

Significant
Findings: Report on questionnaire survey forthcoming in *Final Report.*

Seaver, W. Burleigh, and Arthur H. Patterson. 1976.
               "Decreasing Fuel-Oil Consumption through Feedback and Social
               Commendation." *Journal of Applied Behavior Analysis,* 9 (2):
               147-59.

Subject: Consumer behavioral patterns of fuel consumption.

Experiment
Date: February-May 1974.

Methods: A sample of 180 households was randomly drawn from a local
               fuel-oil distributor in a university community in central Pennsyl-
               vania. To control for differences in home size, insulation, age, and

furnace efficiency, households were assigned to blocks according to their oil consumption in the January-May period of the previous year. Customers were randomly assigned to the three treatment conditions (60 households in each condition). The control group received only the normal delivery ticket (number of gallons delivered, price, and amount due); the feedback group received the normal delivery ticket plus a feedback slip concerning rates of use for the current period and the previous year, percent increase or decrease in consumption rate, and the resulting dollar savings or loss based on this information; the feedback-plus-commendation group received the delivery ticket, feedback information, and a decal ("We Are Saving Oil") if they had reduced their consumption rate compared with the previous winter.

Analysis Techniques:	Analysis of variance on 1974 consumption rates; a multiple comparisons test (using Tukey's WDS technique) was performed on the condition means.

Significant Findings:	The consumption rate for the feedback-plus-reward group was significantly lower than either of the two other groups ($p < 0.01$).

No difference between feedback group and control group; no significant conditions-by-blocks interaction showed that the effectiveness of the feedback plus reward did not depend on the base-line consumption rates of the households.

Authors feel that if rewards for reduced consumption (following oil deliveries) had been more frequent and of more practical utility to consumers, the effect of the procedure might have been greater.

Of the 57 usable households in the feedback-plus-reward condition, 43 (75 percent) reduced their consumption compared with the previous winter.

Socolow, Robert H.; David T. Harrje; Lawrence Mayer; and Clive Seligman. Ongoing.

> *Energy Conservation in Housing.* Princeton, New Jersey: Princeton University, Center for Environmental Studies.

Subject:	Engineering, architectural, behavioral, and social aspects of household consumption.

Experiment Date:	Ongoing since 1971.

Methods: Intensive investigation of a planned development of 3,000 dwelling units. Sets of identical units and appliance packages, highly instrumented and monitored units, mix of single-family, town house, and apartment units available.

Analysis   Analysis of variance on feedback experiments, modeling and time-
Techniques: series explorations, analysis of retrofitting.

Significant   Behavioral and structural features are both important in conserva-
Findings: tion strategy.
                       Immediate feedback of daily electrical use led to decrease in consumption. Authors conclude that feedback is an effective conservation strategy and is more successful with moderate users of electricity than with high users.

Stearns, Mary D. 1975.
                  The Social Impacts of the Energy Shortage: Behavioral and Attitude Shifts. Washington, D.C.: U.S. Department of Transportation.

Subject: Responses to and attitudes toward energy shortages. (Also includes trip-making behavior—not summarized here.)

Survey Date: November, December 1973; February 1974.

Methods: National random sample survey (telephone interviews) by NORC, Continuous National Survey (n = 700).

Analysis
Techniques: Frequencies, cross tabulations.

Significant   High social status related to an abstract understanding of the
Findings: energy problem, including impacts that did not focus on personal effects.
                       Households expressed more tolerance of strict conservation policies before the shortage, and were increasingly less receptive to onerous potential policies during the energy shortage.
                       Overall belief in warnings of impending severity of shortages.
                       Better-educated households viewed energy problem as more important, but reported less severe personal experiences.
                       Older respondents expected the energy problem to be of shorter duration and reported that it had little effect on their lives.

Talarzyk, W. Wayne, and Glenn S. Omura. 1975.
"Consumer Attitudes Toward and Perceptions of the Energy Crisis." In *American Marketing Association 1974 Combined Proceedings*, ed. Ronald C. Curhan, pp. 316-22. Ann Arbor, Michigan: Xerox University Microfilms.

Subject: Attitudinal and behavioral responses toward energy crisis.

Survey Date: March 1974.

Methods: Mail survey; national sample of 1,000 households balanced to parallel census data for geographic divisions; 662 usable responses (772 returned).

Analysis
Techniques: Frequencies, cross tabulations, factor analysis.

Significant   Positive, but not enthusiastic, attitude toward energy conserva-
Findings: tion pleas. Only minor changes in life style.
               Blame—heavy on oil companies (most believe crisis contrived).
               35 percent said crisis greatly reduced their income, and 47 percent said it would get worse before it gets better.
               Older persons had less attitudinal resistance to conservation pleas, but oldest and youngest age groups made least change in their activities.
               Highest income and education groups most likely to report changes in activities.
               No relationship between demographic variables and blame for crisis.

Thompson, Phyllis T., and John Mactavish. 1976.
Energy Problems: Public Beliefs, Attitudes, and Behaviors. Allendale, Michigan: Grand Valley State College, Urban and Environmental Studies Institute.

Subject: Public perceptions and attitudes regarding energy-related problems and conservation behavior.

Survey Date: February 1976.

Methods: Personal interviews with a random sample of the population of Grand Rapids, Michigan (85.8 percent response rate = 515 completed interviews). Deliberately structured to provide half male and half female respondents.

Analysis
Techniques: Cross tabulations, frequency distributions.

Significant   More than 50 percent are cynical about the nature of the problem;
Findings: they do not connect energy problems and diminishing supply.
                      Behavior consistent with beliefs—few or no conservation measures.
                      Group tends to be older, at lower educational and occupational
                      levels, and to rely on TV for their information.
                          Smaller group (20 percent) believes in real and persistent
                      shortages and has adopted a variety of conservation measures.
                      Tends to be under 45, at skilled and professional occupational
                      levels, have college or graduate degrees, and get information from
                      newspapers, national magazines, and research reports.
                          Large group of public is consistently unsure—does not
                      know who or what to believe.
                          Educational program with simple, consistent messages
                      needed, incorporating new approaches that will lead the public
                      to reach appropriate conclusions from their own synthesis of the
                      data provided.

Walker, Nolan E., and E. Linn Draper. 1975.
                      "The Effects of Electricity Price Increases on Residential Usage
                      by Three Economic Groups: A Case Study." In *Texas Nuclear
                      Power Policies* 5. Austin: University of Texas, Center for Energy
                      Studies.

Subject: Impact of price increases on income groups; behavior and atti-
                      tudes; electricity consumption changes.

Survey Date: July 1974; consumption data from previous two years.

Methods: Stratified random sample of households in Austin, Texas; 60
                      personal interviews; electricity consumption data from company.

Analysis
Techniques: Charting of consumption data over time period; frequencies.

Significant   From July 1972 to July 1974, number of lower-income house-
Findings: holds increasing use equaled number decreasing use of electricity;
                      middle-income, number decreasing electricity consumption

greater than number increasing; upper-income, number decreasing much less than number increasing.

Authors conclude that upper-income households will continue to consume regardless of price; lower-income households already at minimum; greatest flexibility in middle-income groups.

Warkov, Seymour. Ongoing.
>*A Case Study of Energy Conservation at Large Public Facilities: The University of Connecticut.* Storrs: University of Connecticut.

Subject: Attitudes and behavior of university housing residents.

Survey Date: Fall, Winter 1974.

Methods: Mail questionnaire to random sample—177 undergraduates in residence; same questionnaire to 400 dormitory residents in 5 dorms selected on various social and engineering criteria; combined observation-interview at 3 different times of residents found in rooms; follow-up in winter of all 3 phases.

Questionnaire mailed to random sample of 200 "professionals": faculty, administrators, graduate assistants; same questionnaire to random sample of 200 "classified" personnel: clerical, maintenance.

Analysis
Techniques: Not yet available.

Significant
Findings: Not yet available.

Warkov, Seymour. 1976.
>*Energy Conservation in the Houston-Galveston Area Complex: 1976. Preliminary Results from the Houston Metropolitan Area Project.* Houston: University of Houston, Institute for Urban Studies.

Subject: Energy conservation practices and attitudes of Houston-area residents.

Survey Date: Spring, Summer 1976.

Method: Random telephone survey of 3,019 metropolitan-area residents.

Analysis
Techniques: Frequencies, cross tabulations.

Significant   75 percent of the respondents reported curtailing use of lights,
Findings: 65 percent curtailed use of air conditioning, 14 percent had
                   insulated their home or apartment during last year.
                        The higher the income of the household, the higher the
                   consumption of energy and the lower the relative proportion of
                   the household budget spent on energy.
                        Reported curtailment of use of electric lights does not
                   differ among income groups. Curtailment of air conditioning is
                   less at higher income levels. The likelihood of adding insulation is
                   directly related to income level.
                        Renters generally conserve less than owners. House renters'
                   behavior is more similar to homeowners' probably because they
                   are more likely to pay their own utility bills than are apartment
                   renters.
                        Transportation data also reported.

Warren, Donald I. 1974.
                   *Individual and Community Effects on Response to the Energy
                   Crisis of Winter, 1974: An Analysis of Survey Findings from
                   Eight Detroit Area Communities.* Ann Arbor: University of
                   Michigan, Institute of Labor and Industrial Relations, Program
                   in Community Effectiveness.

Subject: Responses to and attitudes toward energy crisis of winter 1973-
                   74.

Survey Date: April-June 1974.

Method: Random sample survey (personal interviews) of 766 households
                   in eight Detroit-area communities.

Analysis
Techniques: Frequencies, cross tabulations, rank-order correlations.

Significant   The energy crisis of 1973-74 was perceived as a failure of the
Findings: institutions of American society rather than as a result of the
                   actions of foreign countries.
                        Middle-class households more likely than high-income
                   or low-income families to believe crisis was real and to report
                   they had experienced shortages and cut consumption.

Vast majority of all respondents reported engaging in conservation behavior.

Social setting (type of neighborhood, socioeconomic level, actions of others) more important than attitudes in affecting conservation behavior.

Warren, Donald I., and David L. Clifford. 1975.
Local Neighborhood Social Structure and Response to the Energy Crisis of 1973-74. Ann Arbor: University of Michigan, Institute of Labor and Industrial Relations.

Subject: Conservation behavior in context of neighborhood patterns.

Survey Date: April-June 1974.

Methods: Cross-section random sample (766) of 8 communities in Detroit metropolitan area; personal interviews with household head or spouse.

Analysis   Responses dichotomized and 6-fold typology constructed; means
Techniques: given for responses on various attitudinal and behavioral variables for each neighborhood type.

Significant   Most (58 percent) believe in at least a "somewhat serious"energy
Findings: shortage. Almost all report trying to conserve.

Those who report being bothered by energy shortages save more.

Belief in reality of crisis has little effect on conservation behavior.

Higher income levels are more likely to believe crisis is real and to report more effects.

Conservation increases with income—sharp break at $10,000.

Six neighborhood types listed (from highest to lowest conservation behavior): integral, parochial, diffuse, stepping-stone, transitory, anomic.

Role of attitudes and income as predictor of conservation behavior valid only in context of neighborhood and community variables. Types of conservation efforts vary by type of community. In general, conservation strategies will be more successful the more they derive from the local setting.

Wascoe, Nancy E.; Stuart W. Cook; and Richard Beatty. Ongoing.
The Effects of Fear Appeals Upon Behavioral Intentions Toward

Energy Consumption: A Replication. Boulder: University of Colorado.

Subject: Can fear appeals concerning the energy crisis affect attitudes and behavioral intentions toward energy consumption?

Experiment
Date: February-May 1976.

Methods: Students in the University of Colorado School of Business read 1 of 8 communications, then completed an attitude and behavioral questionnaire. The 8 communications were the orthogonal combinations of 3 2-level factors:
1. Probability—an energy shortage is or is not highly probable in the near future
2. Severity—if a shortage does occur, it will or will not have severe negative effects on our way of life
3. Efficacy—individuals can or cannot be effective in averting a shortage by conservation.
A behavioral measure was also included—whether or not students responded to an invitation to join an energy conservation project sponsored by the University Environmental Council.

Analysis
Technique: 2 x 2 x 2 analysis of variance.

Significant   Students who read the "severe negative effects" communication
Findings: expressed stronger intentions to conserve energy; those who read the "high efficacy" communication also expressed stronger intentions, but only with regard to group conservation activities. The probability factor had no effect. None of the three factors affected actual behavior.

Winett, Richard A., and Michael T. Neitzel. 1975.
"Behavioral Ecology: Contingency Management of Consumer Energy Use." *American Journal of Community Psychology* 3: 123-33.

Subject: Effect of information and monetary incentives on consumption behavior.

Experiment
Date: January-March 1974.

Methods: 31 volunteer households in Lexington, Kentucky, were adminis-
               tered questionnaires by telephone and had consumption of elec-
               tricity and natural gas monitored for two weeks to provide base-
               line data.
               Households were matched on basis of consumption level
               and randomly assigned to one of two groups. The "Information"
               group received a manual outlining conservation procedures and a
               recording sheet for monitoring their own meter readings. The
               "Incentive" group received the same manual and recording sheet,
               plus cash payment ($2-$20/week) contingent on their meeting
               predetermined levels of energy conservation. Experimental condi-
               tions were in effect for three weeks, and follow-up monitoring
               and interviews were conducted.

Analysis
Techniques: Analysis of variance.

Significant   "Incentive" group averaged approximately 15 percent more
Findings: electricity reduction than "Information" group. Difference
               maintained at two-week follow-up.
               Follow-up analysis at two months revealed weaker trend for
               "Incentive" group than "Information" group to reduce consump-
               tion more.
               Best predictor of energy reduction in both groups was
               base-line use.

Wright, Susan Elizabeth. 1975.
               "Public Responses to the Energy Shortage: An Examination of
               Social Class Variables." Unpublished Ph.D. dissertation, Iowa
               State University.

Subject: Relationship between social class and energy shortage perceptions.

Survey Date: Summer 1974.

Methods: Stratified random sample from Des Moines, Iowa; 190 persons
               interviewed.

Analysis
Techniques: Correlational analysis, regression analysis.

Significant   Social status variables, as a set, explain only a small portion of the
Findings: variance in energy shortage perception.

While some significant relationships between social class and energy shortage perception do exist, these are not strong enough to indicate a social class polarization of interests on the energy shortage issue.

Zuiches, James J.* 1975.
> *Energy and the Family*. East Lansing: Michigan State University, Department of Agricultural Economics.

Subject: Relationship between attitudes about the energy crisis and the acceptability of various policies with direct and indirect consequences for family energy consumption.

Survey Date: May, June 1974.

Methods: Modified area probability sample of urban and rural families in housing SMSA; 217 completed interviews (61.5 percent). First stage of a five-year longitudinal study.

3 phases: (1) screen out nonfamilies; (2) deliver questionnaires to husband/wife/oldest child over 12; (3) collect questionnaires and conduct personal interview with one head of household.

Analysis
Techniques: Frequencies, cross tabulations.

Significant
Findings: Respondents were split 50/50 on the reality of the energy crisis (30 percent of the 50 percent who did not consider the crisis real felt it would be real in the near future). "Considerable variation by sex, education, and energy awareness" in favoring the policies presented was found—females more likely to believe energy crisis real, as were urban dwellers. Strong correlations between amount of education and energy awareness, which leads to belief in reality of energy crisis and acceptance of a diversity of policies.

Urban dwellers more supportive of policies presented (females more than males). Low acceptability policies were rationing meat, restricting electricity use, reversing school seasons, and increasing taxes on large families. High acceptability policies were home gardening, increased home food preparation, tax deductions for home insulation and improvements.

The greater the energy awareness of respondent, the more likely that he or she would find most of the 18 policies acceptable (even those policies ranked lowest).

Education (from academics to mass media exposure) indirectly contributes to energy awareness, which affects policy acceptance, so these two factors seem to be more important.

Previous studies have not found sex differences in public opinion, policy, or concern with environmental issues. In this study females were found to be more responsive to reality of crisis and to have higher energy awareness; this led to more acceptance of policies that facilitate energy conservation.

TABLE C.1

Sex—Factor 1

Factor	Sex of the Subject				Row
	Female		Male		Totals
Extent of the Energy Problem	(n)	(%)	(n)	(%)	(n)
Major problem	232	56.3	569	53.4	801
Minor problem	180	43.7	496	46.6	676
Totals	412	100.0	1,065	100.0	1,477

Note: $x^2 = .88$; df = 1; p $< .35$.
Source: All tables in Appendixes C and D have been compiled by the authors.

TABLE C.2

Age—Factor 1

Factor	Age of the Subject										
	<30 Years		30-39 Years		40-49 Years		50-60 Years		> 60 Years		Row Totals
Extent of the Energy Problem	(n)	(%)	(n)	(%)	(n)	(%)	(n)	(%)	(n)	(%)	(n)
Major problem	224	70.4	186	57.9	147	47.6	137	48.1	97	45.8	791
Minor problem	94	29.6	135	42.1	162	52.4	148	51.9	115	54.2	654
Totals	318	100.0	321	100.0	309	100.0	285	100.0	212	100.0	1,445

Note: $x^2 = 51.40$; df = 4; p = .0578.

TABLE C.3

Education—Factor 1

Factor	Education of the Subject								Row Totals
Extent of the Energy Problem	Less Than High School Education		Attended Trade School		Attended College		Attended Graduate School		
	(n)	(%)	(n)	(%)	(n)	(%)	(n)	(%)	(n)
Major problem	21	22.3	98	30.6	342	57.7	343	72.2	804
Minor problem	73	77.7	222	69.4	251	42.3	132	27.8	678
Totals	94	100.0	320	100.0	593	100.0	475	100.0	1,482

Note: x^2 = 175.06; df = 3; p = .00000.

TABLE C.4

Total Family Income—Factor 1

Factor	Total Family Income of the Subject												Row Totals
Extent of the energy problem	<$5,000		$5,000-$9,999		$10,000-$14,999		$15,000-$19,999		$20,000-$24,999		>$25,000		
	(n)	(%)	(n)	(%)	(n)	(%)	(n)	(%)	(n)	(%)	(n)	(%)	(n)
Major problem	63	52.9	133	46.5	169	50.3	159	54.5	126	60.6	136	67.0	786
Minor problem	56	47.1	153	53.5	167	49.7	133	45.5	82	39.4	67	33.2	658
Totals	119	100.0	286	100.0	336	100.0	292	100.0	208	100.0	203	100.0	1,444

Note: x^2 = 25.75; df = 5; p = .0001.

TABLE C.5

Race—Factor 1

Factor	Race of the Subject						Row Totals
Extent of the Energy Problem	White		Black		Mexican-American		
	(n)	(%)	(n)	(%)	(n)	(%)	(n)
Major problem	697	56.8	12	36.4	60	38.0	769
Minor problem	531	43.2	21	63.6	98	62.0	650
Totals	1,228	100.0	33	100.0	158	100.0	1,419

Note: x^2 = 24.22; df = 2; p = .00001.

TABLE C.6

Social-Psychological Variables—Factor 1

	Extent of the Energy Problem		
	Major Problem ($\bar{x}$)	Minor Problem ($\bar{x}$)	F-Ratio
Dogmatism	31.74	29.22	21.54[a]
Conservatism	29.08	26.64	24.45[a]
Political incapability	11.62	9.79	61.55[a]
Discontent with politics	8.27	7.40	19.01[a]
Big business	13.79	12.56	14.12[a]
Government intervention	15.37	14.69	8.24[b]

[a]p < .000.
[b]p ≤ .004.

TABLE C.7

Sex—Factor 2

Factor Present and Long-Term Impacts of the Energy Problem	Sex of the Subject				Row Totals
	Female		Male		
	(n)	(%)	(n)	(%)	(n)
Substantial problem	280	68.3	646	61.7	926
Not substantial problem	130	31.7	401	38.3	531
Totals	410	100.0	1,047	100.0	1,457

Note: x^2 = 5.24; df = 1; p = .0220.

TABLE C.8

Age—Factor 2

Factor Present and Long-Term Impacts of the Energy Problem	Age of the Subject										Row Totals
	<30 Years		30-39 Years		40-49 Years		50-60 Years		> 60 Years		
	(n)	(%)	(n)	(%)	(n)	(%)	(n)	(%)	(n)	(%)	(n)
Substantial problem	224	72.7	219	67.8	204	66.7	164	58.6	101	47.9	912
Not Substantial problem	84	27.3	104	32.2	102	33.3	116	41.4	110	52.1	516
Totals	308	100.0	323	100.0	306	100.0	280	100.0	211	100.0	1,428

Note: x^2 = 40.49; df = 4; p = .0000.

TABLE C.9

Education—Factor 2

Factor	Education of the Subject								
Present and Long-Term Impacts of the Energy Problem	Less Than High School Education		Attended Trade School		Attended College		Attended Graduate School		Row Totals
	(n)	(%)	(n)	(%)	(n)	(%)	(n)	(%)	(n)
Substantial problem	53	55.2	198	65.3	392	64.5	284	62.4	927
Not substantial problem	43	44.8	105	34.7	216	35.5	171	37.6	535
Totals	96	100.0	303	100.0	608	100.0	455	100.0	1,462

Note: $x^2 = 3.76$; df = 3; p = .2883.

TABLE C.10

Total Family Income—Factor 2

Factor	Total Family Income of the Subject												
Present and Long-term Impacts of the Energy Problem	<$5,000		$5,000-$9,999		$10,000-$14,999		$15,000-$19,999		$20,000-$24,999		>$25,000		Row Totals
	(n)	(%)	(n)	(%)	(n)	(%)	(n)	(%)	(n)	(%)	(n)	(%)	(n)
Substantial problem	70	59.3	179	65.6	222	66.3	202	69.9	116	61.1	123	53.2	912
Not substantial problem	48	40.7	94	34.4	113	33.7	87	30.1	74	38.9	108	46.8	524
Totals	118	100.0	273	100.0	335	100.0	289	100.0	190	100.0	231	100.0	1,436

Note: $x^2 = 18.57$; df = 5; p = .0023.

TABLE C.11

Race—Factor 2

Factor	Race of the Subject						
Present and Long-term Impacts of the Energy Problem	White		Black		Mexican-American		Row Totals
	(n)	(%)	(n)	(%)	(n)	(%)	(n)
Substantial problem	758	62.7	19	65.5	98	62.8	875
Not substantial problem	451	37.3	10	34.5	58	37.2	519
Totals	1,209	100.0	29	100.0	156	100.0	1,394

Note: $x^2 = .096$; df = 2; p = .95.

TABLE C.12

Social-Psychological Variables—Factor 2

	Present and Long-Term Impacts of the Energy Problem		
	Substantial	Not Substantial	F-Ratio
Dogmatism	30.55	30.49	.029
Conservatism	27.91	27.35	1.12
Political incapability	10.45	11.39	12.21*
Discontent with politics	7.39	8.7	42.81*
Big business	12.43	15.03	58.81*
Government intervention	14.66	15.79	20.63*

*$p < .000$.

TABLE C.13

Sex—Factor 3

Factor Responsibility for the Energy Problem	Sex of the Subject				Row Totals
	Female		Male		
	(n)	(%)	(n)	(%)	(n)
Willing to assign responsibility	231	52.0	527	46.3	758
Not willing to assign responsibility	213	48.0	611	53.7	824
Totals	444	100.0	1,138	100.0	1,582

Note: $x^2 = 3.95$; df = 1; p = .0467.

TABLE C.14

Age—Factor 3

Factor Responsibility for the Energy Problem	Age of the Subject										Row Totals
	<30 Years		30-39 Years		40-49 Years		50-60 Years		> 60 Years		
	(n)	(%)	(n)	(%)	(n)	(%)	(n)	(%)	(n)	(%)	(n)
Willing to assign responsibility	185	54.4	149	44.0	156	48.0	147	48.2	104	42.4	741
Not willing to assign responsibility	155	45.6	190	56.0	169	52.0	158	51.8	141	57.6	813
Totals	340	100.0	339	100.0	325	100.0	305	100.0	245	100.0	1,554

Note: $x^2 = 10.79$; df = 4; p = .0289.

TABLE C.15

Education—Factor 3

Factor	Education of the Subject								
Responsibility for the Energy Problem	Less Than High School Education		Attended Trade School		Attended College		Attended Graduate School		Row Totals
	(n)	(%)	(n)	(%)	(n)	(%)	(n)	(%)	(n)
Willing to assign responsibility	71	67.6	195	60.0	319	49.3	175	34.2	760
Not willing to assign responsibility	34	32.4	130	40.0	328	50.7	337	65.8	829
Totals	105	100.0	325	100.0	647	100.0	512	100.0	1,589

Note: $x^2 = 74.56$; df = 3; p < .000.

TABLE C.16

Total Family Income—Factor 3

Factor												
Responsibility for the Energy Problem	Total Family Income of the Subject											Row Totals
	<$5,000		$5,000-$9,999		$10,000-$14,999		$15,000-$19,999		$20,000-$24,999		>$25,000	
	(n)	(%)	(n)	(%)	(n)	(%)	(n)	(%)	(n)	(%)	(n) (%)	(n)
Willing to assign responsibility	71	59.2	165	62.0	184	51.4	148	44.3	93	42.5	83 32.7	744
Not willing to assign responsibility	49	40.8	101	38.0	174	48.6	186	55.7	126	57.5	171 67.3	807
Totals	120	100.0	266	100.0	358	100.0	334	100.0	219	100.0	254 100.0	1,551

Note: $x^2 = 57.03$; df = 5; p = .00001.

TABLE C.17

Race—Factor 3

Factor	Race of the Subject						
Responsibility for the Energy Problem	White		Black		Mexican-American		Row Totals
	(n)	(%)	(n)	(%)	(n)	(%)	(n)
Willing to assign responsibility	594	45.0	20	60.6	113	66.1	727
Not willing to assign responsibility	726	55.0	13	39.4	58	33.9	797
Totals	1,320	100.0	33	100.0	171	100.0	1,524

Note: x^2 = 29.22; df = 2; p < .000.

TABLE C.18

Social-Psychological Variables—Factor 3

	Responsibility for the Energy Problem		
	Willing to Assign Responsibility	Not Willing to Assign Responsibility	F-Ratio
Dogmatism	28.80	32.42	46.76*
Conservatism	26.56	29.00	29.53*
Political incapability	9.61	12.2	132.49*
Discontent with politics	7.07	8.96	105.49*
Big business	11.47	15.56	175.64*
Government intervention	13.98	16.40	120.39*

*p < .000.

TABLE C.19

Sex—Factor 4

Factor	Sex of the Subject				Row Totals
	Female		Male		
Corporation Practices	(n)	(%)	(n)	(%)	(n)
Positive attitude	203	48.8	502	44.7	705
Negative attitude	213	51.2	622	55.3	835
Totals	416	100.0	1,124	100.0	1,540

Note: $x^2 = 1.93$; df = 1; p = .164.

TABLE C.20

Age—Factor 4

Factor Corporation Practices	Age of the Subject										Row Totals
	<30 Years		30-39 Years		40-49 Years		50-60 Years		> 60 Years		
	(n)	(%)	(n)	(%)	(n)	(%)	(n)	(%)	(n)	(%)	(n)
Positive attitude	111	37.5	123	37.7	122	38.9	164	52.6	160	62.7	680
Negative attitude	185	62.5	203	62.3	192	61.1	148	47.4	95	37.3	823
Totals	296	100.0	326	100.0	314	100.0	312	100.0	255	100.0	1,503

Note: $x^2 = 58.04$; df = 4; p = .00001.

TABLE C.21

Education—Factor 4

Factor	Education of the Subject								
Corporation Practices	Less Than High School Education		Attended Trade School		Attended College		Attended Graduate School		Row Totals
	(n)	(%)	(n)	(%)	(n)	(%)	(n)	(%)	(n)
Positive attitude	52	51.0	139	40.1	252	41.7	261	53.3	704
Negative attitude	50	49.0	208	59.9	353	58.3	229	46.7	840
Totals	102	100.0	347	100.0	605	100.0	490	100.0	1,544

Note: x^2 = 20.89; df = 3; p = .0001.

TABLE C.22

Total Family Income—Factor 4

Factor	Total Family Income of the Subject												
Corporation Practices	<$5,000		$5,000-$9,999		$10,000-$14,999		$15,000-$19,999		$20,000-$24,999		>$25,000		Row Totals
	(n)	(%)	(n)	(%)	(n)	(%)	(n)	(%)	(n)	(%)	(n)	(%)	(n)
Positive attitude	67	54.9	116	40.7	133	40.8	123	40.7	100	44.2	141	57.6	680
Negative attitude	55	45.1	169	59.3	193	59.2	179	59.3	126	55.8	104	42.4	826
Totals	122	100.0	285	100.0	326	100.0	302	100.0	226	100.0	245	100.0	1,506

Note: x^2 = 27.14; df = 5; p = .0001.

TABLE C.23

Race—Factor 4

Factor	Race of the Subject						Row Totals
Corporation Practices	White		Black		Mexican-American		
	(n)	(%)	(n)	(%)	(n)	(%)	(n)
Positive attitude	593	46.3	16	48.5	72	40.9	681
Negative attitude	688	53.7	17	51.5	104	59.1	809
Totals	1,281	100.0	33	100.0	176	100.0	1,490

Note: x^2 = 1.9; df = 2; p = .384.

172

TABLE C.24

Social-Psychological Variables—Factor 4

	Corporation Practices		
	Positive-Neutral Attitude	Negative Attitude	F-Ratio
Dogmatism	29.97	30.78	2.28
Conservatism	26.94	28.13	6.21[a]
Political incapability	11.37	10.39	17.87[b]
Discontent with politics	8.63	7.51	32.96[b]
Big business	16.05	11.47	214.16[b]
Government intervention	16.10	14.38	53.65[b]

[a] $p < .02$.
[b] $p < .000$.

TABLE C.25

Sex—Factor 5

Factor Who Has Taken Advantage of the Energy Problem	Sex of the Subject				Row Totals
	Female		Male		
	(n)	(%)	(n)	(%)	(n)
Energy companies have taken advantage	344	74.5	863	71.4	1,207
Energy companies have not taken advantage	118	25.5	346	28.6	464
Totals	462	100.0	1,209	100.0	1,671

Note: $x^2 = 1.42$; df = 1; $p < .24$.

Table C.26

Age—Factor 5

Factor	Age of the Subject									
Who Has Taken Advantage of the Energy Problem	<30 Years		30-39 Years		40-49 Years		50-60 Years		>60 Years	Row Totals
	(n)	(%)	(n)	(%)	(n)	(%)	(n)	(%)	(n)	(%)	(n)
Energy companies have taken advantage	247	73.3	261	72.9	259	73.8	239	71.6	171	66.5	1,177
Energy companies have not taken advantage	90	26.7	97	27.1	92	26.2	95	28.4	86	33.5	460
Totals	337	100.0	358	100.0	351	100.0	334	100.0	257	100.0	1,637

Note: $x^2 = 4.8$; df = 4; p = .3083.

TABLE C.27

Education—Factor 5

Factor	Education of the Subject								
Who Has Taken Advantage of the Energy Problem	Less Than High School Education		Attended Trade School		Attended College		Attended Graduate School		Row Totals
	(n)	(%)	(n)	(%)	(n)	(%)	(n)	(%)	(n)
Energy companies have taken advantage	90	81.1	275	74.1	492	71.9	351	69.1	1,208
Energy companies have not taken advantage	21	18.9	96	25.9	192	28.1	157	30.9	466
Totals	111	100.0	371	100.0	684	100.0	508	100.0	1,674

Note: $x^2 = 7.5$; df = 3; p = .0574.

TABLE C.28

Total Family Income—Factor 5

Factor Who Has Taken Advantage of the Energy Crisis	Total Family Income of the Subject												Row Totals
	<$5,000		$5,000-$9,999		$10,000-$14,999		$15,000-$19,999		$20,000-$24,999		> $25,000		
	(n)	(%)	(n)	(%)	(n)	(%)	(n)	(%)	(n)	(%)	(n)	(%)	(n)
Energy companies have taken advantage	102	76.7	211	71.8	292	75.6	243	72.8	171	72.5	168	66.9	1,187
Energy companies have not taken advantage	31	23.3	83	28.2	94	24.4	91	27.2	65	27.5	83	33.1	447
Totals	133	100.0	294	100.0	386	100.0	334	100.0	236	100.0	251	100.0	1,634

Note: $x^2 = 7.08$; df = 5; p = .2141.

TABLE C.29

Race—Factor 5

Factor Who Has Taken Advantage of the Energy Problem	Race of the Subject						Row Totals
	White		Black		Mexican-American		
	(n)	(%)	(n)	(%)	(n)	(%)	(n)
Energy companies have taken advantage	988	71.2	28	77.8	138	78.9	1,154
Energy companies have not taken advantage	400	28.8	8	22.2	37	21.1	445
Totals	1,388	100.0	36	100.0	175	100.0	1,599

Note: $x^2 = 5.13$; df = 2; p = .0767.

TABLE C.30

Social-Psychological Variables—Factor 5

	Who Has Taken Advantage of the Energy Problem		
	Energy Companies Have Taken Advantage of the Problem	Energy Companies Have Not Taken Advantage of the Problem	
	$\overline{x}$	$\overline{x}$	F-Ratio
Dogmatism	30.32	31.26	2.72
Conservatism	27.78	28.07	.32
Political incapability	10.67	11.34	7.42[a]
Discontent with politics	7.77	8.5	13.63[b]
Big business	12.97	14.97	32.94[b]
Government intervention	14.79	16.09	26.98[b]

[a]$p < .01$.
[b]$p < .000$.

TABLE C.31

Sex—Factor 6

Factor	Sex of the Subject				Row
Attempts to Solve the Energy Problem	Female		Male		Totals
	(n)	(%)	(n)	(%)	(n)
Substantial attempts	170	45.6	401	39.6	571
Not substantial attempts	203	54.4	611	60.4	814
Totals	373	100.0	1,012	100.0	1,385

Note: $x^2 = 3.74$; df = 1; $p = .0530$.

TABLE C.32

Age—Factor 6

Factor	Age of the Subject										
Attempts to Solve the Energy Problem	<30 Years		30-39 Years		40-49 Years		50-60 Years		> 60 Years		Row Totals
	(n)	(%)	(n)	(%)	(n)	(%)	(n)	(%)	(n)	(%)	(n)
Substantial attempts	177	26.3	94	31.2	104	38.1	115	44.4	157	71.4	547
Not substan- attempts	216	73.7	207	68.8	169	61.9	144	55.6	63	28.6	799
Totals	293	100.0	301	100.0	273	100.0	259	100.0	220	100.0	1,346

Note: x^2 = 124.43; df = 4; p = .0000.

TABLE C.33

Education—Factor 6

Factor	Education of the Subject								
Attempts to Solve the Energy Problem	Less Than High School Education		Attended Trade School		Attended College		Attended Graduate School		Row Totals
	(n)	(%)	(n)	(%)	(n)	(%)	(n)	(%)	(n)
Substantial attempts	73	67.0	135	46.9	191	34.8	174	39.5	573
Not substantial attempts	36	33.0	153	53.1	358	65.2	267	60.5	814
Totals	109	100.0	288	100.0	549	100.0	441	100.0	1,387

Note: x^2 = 43.53; df = 3; p = .0000.

TABLE C.34

Total Family Income—Factor 6

Factor						Total Family Income of the Subject							
Attempts to Solve the Energy Problem	<$5,000		$5,000-$9,999		$10,000-$14,999		$15,000-$19,999		$20,000-$24,999		>$25,000		Row Totals
	(n)	(%)	(n)	(%)	(n)	(%)	(n)	(%)	(n)	(%)	(n)	(%)	(n)
Substantial attempts	65	59.1	111	43.5	135	40.7	99	37.2	61	32.6	80	39.8	551
Not substantial attempts	45	40.9	144	56.5	197	59.3	167	62.8	126	67.4	121	60.2	800
Totals	110	100.0	255	100.0	332	100.0	266	100.0	187	100.0	201	100.0	1,351

Note: $x^2 = 22.70$; df = 5; p = .0004.

TABLE C.35

Race—Factor 6

Factor							Row Totals
Attempts to Solve the Energy Problem	White		Black		Mexican-American		
	(n)	(%)	(n)	(%)	(n)	(%)	(n)
Substantial attempts	460	40.0	17	51.5	77	50.7	554
Not substantial attempts	689	60.0	16	48.5	75	49.3	780
Totals	1,149	100.0	33	100.0	152	100.0	1,334

Note: $x^2 = 7.62$; df = 2; p = .0221.

TABLE C.36

Social-Psychological Variables—Factor 6

	Attempts to Solve the Energy Problem		
	Substantial Attempts	Not Substantial Attempts	F-Ratio
Dogmatism	28.34	32.03	42.23[b]
Conservatism	25.62	29.45	53.99[b]
Political incapability	10.57	10.80	1.31
Discontent with politics	8.35	7.55	15.50[b]
Big business	15.46	11.87	113.79[b]
Government intervention	15.30	14.31	3.91[a]

[a] $p < .05$.
[b] $p < .000$.

DEMOGRAPHIC, SOCIAL-PSYCHOLOGICAL, AND COMMUNICATION PATTERNS FOR CONSERVATION FACTORS 1-4

TABLE D.1

Total Family Income—Factor 1

Factor Adjusted Thermostat and Lighting	Total Family Income of the Subject											
	<$5,000		$5,000-$9,999		$10,000-$14,999		$15,000-$19,999		$20,000-$24,999		>$25,000	Row Totals
	(n)	(%)	(n)	(%)	(n)	(%)	(n)	(%)	(n)	(%)	(n)	(%)	(n)
More conserving	170	87.6	362	87.4	455	84.1	399	84.0	258	79.4	279	78.6	1,923
Less conserving	24	12.4	52	12.6	86	15.9	76	16.0	67	20.6	76	21.4	381
Totals	194	100.0	414	100.0	541	100.0	475	100.0	325	100.0	355	100.0	2,304

Note: $x^2 = 17.46$; df = 5; p < .004.

TABLE D.2

Age—Factor 1

Factor Adjusted Thermostat and Lighting	Age of the Subject										
	<30 Years		30-39 Years		40-49 Years		50-60 Years		>60 Years	Row Totals
	(n)	(%)	(n)	(%)	(n)	(%)	(n)	(%)	(n)	(%)	(n)
More conserving	404	84.5	409	82.1	391	82.0	395	84.0	325	85.5	1,925
Less conserving	74	15.5	89	17.9	86	18.0	75	16.0	55	14.5	379
Totals	478	100.0	498	100.0	477	100.0	470	100.0	380	100.0	2,304

Note: $x^2 = 3.08$; df = 4; p < .54.

TABLE D.3

Education—Factor 1

Factor	Education of the Subject								Row
Adjusted Thermostat and Lighting	Less Than High School Education		Attended Trade School		Attended College		Attended Graduate School		Row Totals
	(n)	(%)	(n)	(%)	(n)	(%)	(n)	(%)	(n)
More conserving	148	88.6	431	83.2	807	84.5	588	81.4	1,974
Less conserving	19	11.4	87	16.8	148	15.5	134	18.6	388
Totals	167	100.0	518	100.0	955	100.0	722	100.0	2,362

Note: $x^2 = 6.14$; df = 3; p < .11.

TABLE D.4

Sex—Factor 1

Factor	Sex of the Subject				Row
Adjusted Thermostat and Lighting	Female		Male		Totals
	(n)	(%)	(n)	(%)	(n)
More conserving	578	88.0	1,388	81.7	1,966
Less conserving	79	12.0	310	18.3	389
Totals	657	100.0	1,698	100.0	2,355

Note: $x^2 = 12.89$; df = 1; p < .000.

TABLE D.5

Race—Factor 1

Factor	Race of the Subject						Row
Adjusted Thermostat and Lighting	White		Black		Mexican-American		Row Totals
	(n)	(%)	(n)	(%)	(n)	(%)	(n)
More consuming	1,649	84.1	49	87.5	202	82.1	1,900
Less consuming	312	15.9	7	12.5	44	17.9	363
Totals	1,961	100.0	56	100.0	246	100.0	2,263

Note: $x^2 = 1.16$; df = 2; p < .56.

TABLE D.6

Social-Psychological Variables and Communication Patterns—
Factor 1, Adjusted Thermostat and Lighting

	More Conserving ($\bar{x}$)	Less Conserving ($\bar{x}$)	F-Ratio
Social-psychological variables			
Dogmatism	30.41	30.46	.008
Conservatism	27.72	27.16	1.10
Political incapability	10.81	10.91	.143
Political discontent	7.96	8.14	.695
Big business	13.46	14.18	4.27[a]
Big government	15.08	15.63	4.77[a]
Communication patterns			
Spouse	1.66	2.03	35.28[c]
Children	2.15	2.59	30.23[c]
Friends	2.08	2.32	24.85[c]
People at work	1.93	2.22	22.29[c]
Neighbors	2.54	2.85	24.33[c]
Complainers' index	8.39	7.89	5.40[b]

[a] $p < .05$.
[b] $p < .02$.
[c] $p < .000$.

TABLE D.7

Total Family Income—Factor 2

Factor	Total Family Income of the Subject											
Improved Use Patterns	<$5,000		$5,000-$9,999		$10,000-$14,999		$15,000-$19,999		$20,000-$24,999		>$25,000	Row Totals
	(n)	(%)	(n)	(%)	(n)	(%)	(n)	(%)	(n)	(%)	(n) (%)	(n)
More conserving	164	84.5	334	80.7	403	74.5	315	66.3	183	56.3	180 50.7	1,577
Less conserving	30	15.5	80	19.3	138	25.5	160	33.7	142	43.7	175 49.3	725
Totals	194	100.0	414	100.0	541	100.0	475	100.0	325	100.0	355 100.0	2,302

Note: $x^2 = 136.18$; df = 5; $p < .000$.

TABLE D.8

Age—Factor 2

Factor	Age of the Subject										Row Totals
Improved Use Patterns	<30 Years		30-39 Years		40-49 Years		50-60 Years		> 60 Years		
	(n)	(%)	(n)	(%)	(n)	(%)	(n)	(%)	(n)	(%)	(n)
More conserving	384	80.3	329	66.1	320	67.1	283	60.2	260	68.4	1,576
Less conserving	94	19.7	169	33.9	157	32.9	187	39.8	120	31.6	727
Totals	478	100.0	498	100.0	477	100.0	470	100.0	380	100.0	2,303

Note: $x^2 = 47.73$; df = 4; p < .000.

TABLE D.9

Education—Factor 2

Factor	Education of the Subject								Row Totals
Improved Use Patterns	Less Than High School Education		Attended Trade School		Attended College		Attended Graduate School		
	(n)	(%)	(n)	(%)	(n)	(%)	(n)	(%)	(n)
More conserving	134	80.2	391	75.5	655	68.6	440	60.9	1,620
Less conserving	33	19.8	127	24.5	300	31.4	282	39.1	742
Totals	167	100.0	518	100.0	955	100.0	722	100.0	2,362

Note: $x^2 = 41.54$; df = 3; p < .00000.

TABLE D.10

Sex—Factor 2

Factor	Sex of the Subject				Row Totals
	Female		Male		
Improved Use Patterns	(n)	(%)	(n)	(%)	(n)
More conserving	507	77.2	1,109	65.3	1,616
Less conserving	150	22.8	589	34.7	739
Totals	657	100.0	1,698	100.0	2,355

Note: $x^2 = 30.77$; df = 1; p < .000.

TABLE D.11

Race—Factor 2

Factor	Race of the Subject						Row Totals
Improved Use Patterns	White		Black		Mexican-American		
	(n)	(%)	(n)	(%)	(n)	(%)	(n)
More conserving	1,305	66.5	45	80.4	200	81.3	1,550
Less conserving	656	33.5	11	19.6	46	18.7	713
Totals	1,961	100.0	56	100.0	246	100.0	2,263

Note: $x^2 = 25.79$; df = 2; p < .000.

TABLE D.12

Social-Psychological Variables and Communication Patterns— Factor 2, Improved Use Patterns

	More Conserving $(\bar{x})$	Less Conserving $(\bar{x})$	F-Ratio
Social-psychological variables			
Dogmatism	29.88	31.60	13.66[a]
Conservatism	27.29	28.35	6.23[b]
Political incapability	10.46	11.63	33.47[a]
Political discontent	7.81	8.40	12.67[a]
Big business	13.06	14.71	35.07[a]
Big government	14.67	16.28	64.36[a]
Communication patterns			
Spouse	1.60	1.98	56.77[a]
Children	2.14	2.41	17.02[a]
Friends	2.04	2.29	42.66[a]
People at work	1.90	2.15	26.30[a]
Neighbors	2.49	2.82	41.92[a]
Complainers' index	8.48	7.94	9.65[b]

[a]p < .02.
[b]p < .000.

TABLE D.13

Total Family Income—Factor 3

Factor Reduced Energy-Consuming Activities	Total Family Income of the Subject											
	<$5,000 (n)	(%)	$5,000-$9,999 (n)	(%)	$10,000-$14,999 (n)	(%)	$15,000-$19,999 (n)	(%)	$20,000-$24,999 (n)	(%)	>$25,000 (n)	(%)	Row Totals (n)
More consuming	71	36.6	102	24.6	98	18.1	65	13.7	32	9.8	29	8.2	397
Less consuming	123	63.4	312	75.4	443	81.9	410	86.3	293	90.3	326	91.8	1,907
Totals	194	100.0	414	100.0	541	100.0	475	100.0	325	100.0	355	100.0	2,304

Note: x^2 = 104.29; df = 5; p < .000.

TABLE D.14

Age—Factor 3

Factor Reduced Energy-Consuming Activities	Age of the Subject										
	<30 Years (n)	(%)	30-39 Years (n)	(%)	40-49 Years (n)	(%)	50-60 Years (n)	(%)	>60 Years (n)	(%)	Row Totals (n)
More conserving	53	11.1	60	12.0	86	18.0	92	19.6	104	27.4	395
Less conserving	425	88.9	438	88.0	391	82.0	378	80.4	276	72.6	1,908
Totals	478	100.0	498	100.0	477	100.0	470	100.0	380	100.0	2,303

Note: x^2 = 51.61; df = 4; p < .001.

TABLE D.15

Education—Factor 3

Factor	Education of the Subject								
Reduced Energy-Consuming Activities	Less Than High School Education		Attended Trade School		Attended College		Attended Graduate School		Row Totals
	(n)	(%)	(n)	(%)	(n)	(%)	(n)	(%)	(n)
More conserving	63	37.7	133	25.7	121	12.7	95	13.2	412
Less conserving	104	62.3	385	74.3	834	87.3	627	86.8	1,950
Totals	167	100.0	518	100.0	955	100.0	722	100.0	2,362

Note: x^2 = 96.39; df = 3; p < .00000.

TABLE D.16

Sex—Factor 3

Factor	Sex of the Subject				Row Totals
Reduced Energy-Consuming Activities	Female		Male		
	(n)	(%)	(n)	(%)	(n)
More conserving	212	32.3	421	24.8	633
Less conserving	445	67.7	1,277	75.2	1,722
Totals	657	100.0	1,698	100.0	2,355

Note: x^2 = 13.08; df = 1; p < .000.

TABLE D.17

Race—Factor 3

Factor	Race of the Subject						Row Totals
Reduced Energy-Consuming Activities	White		Black		Mexican-American		
	(n)	(%)	(n)	(%)	(n)	(%)	(n)
More conserving	316	16.1	18	32.1	64	26.0	398
Less conserving	1,645	83.9	38	67.9	182	74.0	1,865
Totals	1,961	100.0	56	100.0	246	100.0	2,263

Note: x^2 = 23.17; df = 2; p < .000.

TABLE D.18

Social-Psychological Variables and Communication Patterns—Factor 3, Reduced Energy-Consuming Activities

	More Conserving ($\bar{x}$)	Less Conserving ($\bar{x}$)	F-Ratio
Social-psychological variables			
Dogmatism	27.40	31.05	40.49*
Conservatism	25.10	28.17	35.46*
Political incapability	9.60	11.08	35.50*
Political discontent	7.32	8.14	16.03*
Big business	12.22	13.87	23.40*
Big government	13.45	15.54	73.42*
Communication patterns			
Spouse	1.31	1.81	70.25*
Children	1.62	2.35	88.67*
Friends	1.80	2.18	65.56*
People at work	1.50	2.08	85.85*
Neighbors	2.00	2.72	143.37*
Complainers' index	8.98	8.17	15.09*

*p < .000.

TABLE D.19

Total Family Income—Factor 4

Factor	Total Family Income of the Subject												
Installed Energy-Conserving Materials	<$5,000		$5,000-$9,999		$10,000-$14,999		$15,000-$19,999		$20,000-$24,999		>$25,000		Row Totals
	(n)	(%)	(n)	(%)	(n)	(%)	(n)	(%)	(n)	(%)	(n)	(%)	(n)
More conserving	94	48.5	154	37.2	149	27.5	91	19.2	57	17.5	67	18.9	612
Less conserving	100	51.5	260	62.8	392	72.5	384	80.0	268	82.5	288	81.1	1,692
Totals	194	100.0	414	100.0	541	100.0	475	100.0	325	100.0	355	100.0	2,304

Note: x^2 = 109.61; df = 5; p < .000.

TABLE D.20

Age—Factor 4

Factor Installed Energy-Conserving Materials	Age of Subject										Row Totals
	<30 Years		30-39 Years		40-49 Years		50-60 Years		>60 Years		
	(n)	(%)	(n)	(%)	(n)	(%)	(n)	(%)	(n)	(%)	(n)
More conserving	132	27.6	97	19.5	119	24.9	107	22.8	146	38.4	601
Less conserving	346	72.4	401	80.5	358	75.1	363	77.2	234	61.6	1,702
Totals	478	100.0	498	100.0	477	100.0	470	100.0	380	100.0	2,303

Note: $x^2 = 44.48$; df = 4; p < .000.

TABLE D.21

Education—Factor 4

Factor Installed Energy-Conserving Materials	Education of the Subject								Row Totals
	Less Than High School Education		Attended Trade School		Attended College		Attended Graduate School		
	(n)	(%)	(n)	(%)	(n)	(%)	(n)	(%)	(n)
More conserving	73	43.7	173	33.4	235	24.6	149	20.6	630
Less conserving	94	56.3	345	66.6	720	75.4	573	79.4	1,732
Totals	167	100.0	518	100.0	955	100.0	722	100.0	2,362

Note: $x^2 = 52.3$; df = 3; p < .000.

TABLE D.22

Sex—Factor 4

Factor Installed Energy-Conserving Materials	Sex of the Subject				Row Totals
	Female		Male		
	(n)	(%)	(n)	(%)	(n)
More conserving	212	32.3	421	24.8	633
Less conserving	445	67.7	1,277	75.2	1,722
Totals	657	100.0	1,698	100.0	2,355

Note: $x^2 = 13.08$; df = 1; p < .000.

TABLE D.23

Race—Factor 4

Factor	Race of the Subject						
Installed Energy-Conserving Materials	White		Black		Mexican-American		Row Totals
	(n)	(%)	(n)	(%)	(n)	(%)	(n)
More conserving	490	25.0	21	37.5	92	37.4	603
Less conserving	1,471	75.0	35	62.5	154	62.6	1,660
Totals	1,961	100.0	56	100.0	246	100.0	2,292

Note: $x^2 = 20.68$; df = 2; p < .000.

TABLE D.24

Social-Psychological Variables and Communication Patterns—Factor 4, Installed Energy-Conserving Materials

	More Conserving ($\bar{x}$)	Less Conserving ($\bar{x}$)	F-Ratio
Social-psychological variables			
Dogmatism	28.32	31.19	34.70[b]
Conservatism	26.11	28.18	21.79[b]
Political incapability	9.96	11.14	31.59[b]
Political discontent	7.35	8.23	25.08[b]
Big business	12.88	13.83	10.48[a]
Big government	14.33	15.48	29.51[b]
Communication patterns			
Spouse	1.45	1.82	64.17[b]
Children	1.84	2.36	60.84[b]
Friends	1.96	2.17	27.58[b]
People at work	1.76	2.06	32.68[b]
Neighbors	2.29	2.71	63.40[b]
Complainers' index	8.30	8.31	.004

[a]p < .001.
[b]p < .000.

WILLIAM H. CUNNINGHAM is Associate Professor of Marketing Administration and Associate Dean for Graduate Programs at the College of Business Administration at the University of Texas at Austin, where he is also Coordinator for Commercial Studies at the Center for Energy Studies.

Dr. Cunningham has written numerous journal articles dealing with the topics of international marketing, marketing management, and the legal and social aspects of marketing. In addition, he has published two research monographs with the Bureau of Business Research at the University of Texas at Austin and two marketing textbooks.

Professor Cunningham received his M.B.A. and Ph.D. degrees in marketing administration from Michigan State University.

SALLY COOK LOPREATO is Director of Social Systems Analysis at the Center for Energy Studies, the University of Texas at Austin.

Dr. Lopreato received her Ph.D. degree in sociology from the University of Texas at Austin and did a year of postdoctoral research in Italy before joining the Center. She has published several articles in the areas of political sociology, energy policy, and sociocultural aspects of energy developments.

ALTERNATIVE ENERGY STRATEGIES: Constraints and Opportunities
John Hagel III

THE ENERGY CRISIS AND THE ENVIRONMENT: An International
Perspective
edited by Donald R. Kelley

ENVIRONMENTAL LEGISLATION: A Sourcebook
edited by Mary Robinson Sive

MANAGING ENVIRONMENTAL CHANGE: A Legal and Behavioral
Perspective
Joseph F. DiMento

PERSPECTIVES ON U.S. ENERGY POLICY: A Critique of Regulation
edited by Edward J. Mitchell

*PLANNING AND CONSERVATION: The Emergence of the Frugal Society
Peter W. House and
Edward R. Williams

*THE SUSTAINABLE SOCIETY: Implications for Limited Growth
edited by Dennis Clark Pirages

*Also available in paperback as a PSS Student Edition